ATLAS OF BORDERS

To Ethel, Roman, Octave, Arthur,
Baptiste, Louis, Thomas and Martin.

DELPHINE PAPIN
BRUNO TERTRAIS

ATLAS OF BORDERS

MAPS
XEMARTIN LABORDE

CONTENTS

II
BORDERS AT SEA
41

III
BARRIERS AND BORDERS
67

IV
BORDERS OF INTEREST
97

V
DISPUTED BORDERS
123

VI
AT THE EDGES OF EMPIRES
151

CONCLUSION
THE FUTURE OF BORDERS
177

GLOSSARY
189

INTRODUCTION
THE RESURGENCE OF BORDERS

Over the last fifteen years or so, borders have never been out of the news. Brexit, terrorism, migration and refugees, Russia's conflicts, wars in the Middle East, rising tensions in Asia, pandemics. Rarely, in the modern era, has there been so much discourse about borders. The subject is so prevalent that it is central to several recent TV series: the successful Danish/Swedish crime drama *Broen/Bron* (*The Bridge*, 2011) was remade to feature other international borders: US–Mexico, Estonia–Russia, Greece–Türkiye, Malaysia–Singapore, Germany–Austria (as *Der Pass*), and France–UK (as *The Tunnel*). Borders are a source of fascination and lie at the very heart of contemporary geopolitics.

What is a border?

It is a geographical boundary that reflects the relationship between two communities: these relationships may be established by military force or diplomacy, or instead by tradition or cooperation. In a way, it is history inscribed in geography, 'time inscribed in space' (Michel Foucher), or even a 'political isobar' (Jacques Ancel). A border may be a stretch of land, a line or a single point. A border that stretches over a wider area is a place of conquest: a frontier. A linear border separates two spaces. A single-point border could be, for example, an airport.

This book focuses mainly on geographical, land and maritime borders between countries. Contemporary borders are inextricably linked to the modern nation-state, which traces its origins back to the Peace of Westphalia (1648). Borders, theoretically, mark the point where a state's territorial jurisdiction ends. Therefore, they are first and foremost a concept in international law. The International Court of Justice defines a border as the 'exact line or lines where the extension in space of the sovereign powers and rights' of one state meets those of another. The border is wrapped around the edges of the state: an interface, but also a membrane.

Michel Foucher cointed the term 'horogenesis' to describe the creation of borders, but how does this process work in practice? Most often, borders are drawn in blood. Even if we don't take decolonization into account, more than a hundred states have established their borders through war. Only fifty or so countries have been created by peaceful secession. It is very rare for the line of a border to be established by a public referendum, although it can occur, as with Germany–Denmark in 1920.

Historically, borders were created to protect a population, separate armies and people, or divide up territories, as we can see from a few historical examples: 1) the Limes Germanicus was created by the Roman Empire to protect itself from the 'barbarians'; 2) Africa's imperial borders were the result of land being shared out between colonial powers at the Berlin Conference in 1884; 3) the border between North and South Korean was created after the Korean War in the 1950s, and separates two political systems.

More specifically, borders can be created by: 1) the end of a conventional war between states: a peace treaty or act of surrender (an armistice or ceasefire only confirms the de facto borders); 2) territorial expansion (annexation) or imperial expansion (colonialization); 3) independence (decolonization or

The Rhaetian Limes in the late 1st century CE

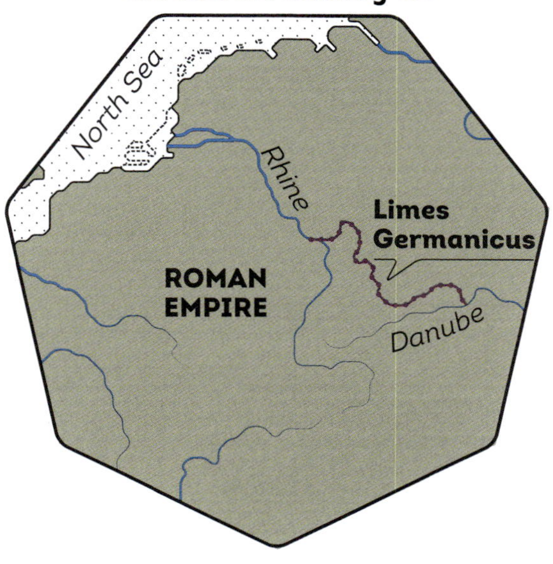

The Wakhan Corridor

The colonial borders of Africa (1884)

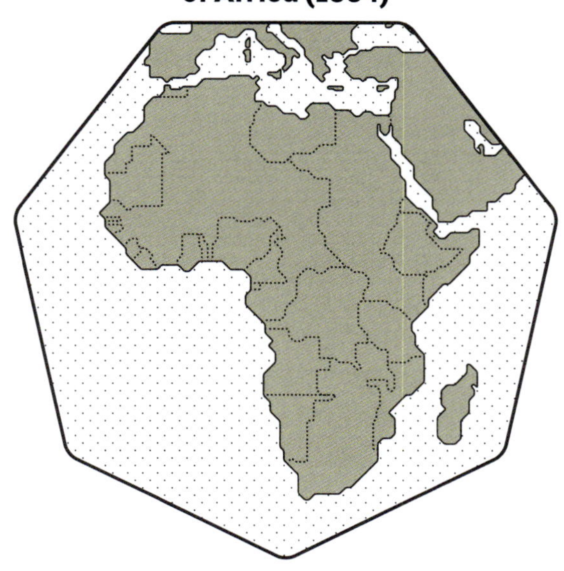

The Jordan-Saudi Arabia border: 'Winston's Hiccup'

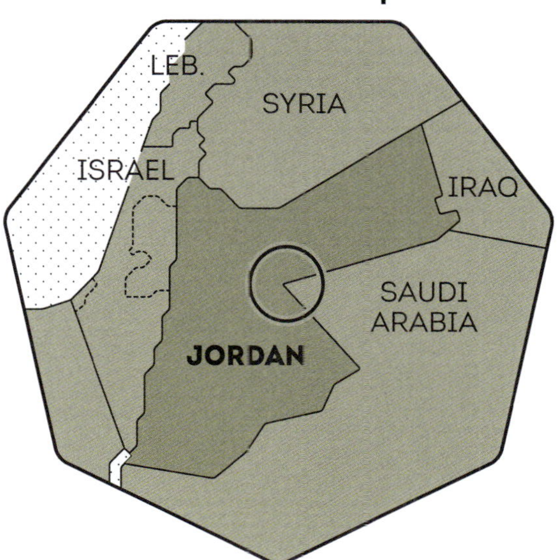

Historically, borders were created to protect a population, separate armies and people, or divide up territories.

secession); 4) amicable cooperation, with borders often being shaped by dynastic claims or land ownership; 5) international arbitration.

While they separate states, it is rare for borders to neatly separate nations, despite the principles put forward by US President Woodrow Wilson after the First World War. The dividing up of imperial spoils at the Paris Peace Conference of 1919 aptly illustrates the range of factors that can be involved when drawing up borders: economic interests (mines, access to the sea), strategic needs (the ability to defend the state), ethnic or national unity or historical tradition.

As for maritime borders, they are often arbitrarily defined at a specific distance from the land, but also take geology (the continental shelf) and hydrology into account.

The British and French were the most prolific creators of borders in the colonial era, driven by their differing priorities: defence for the British, administration for the French. Lord Curzon of Kedleston was probably responsible for creating the most miles of borders (9,600 km). In 1990, more than half of existing land borders traced their origins back to colonial times. These borders imposed by outsiders and bureaucrats, drawn up by empires, have sometimes resulted in strange shapes designed to prevent contact between colonial possessions (the Caprivi Strip in Namibia, the Wakhan Corridor in Afghanistan) or to separate communities (the Galilee panhandle in Israel). Some borders look so strange that we cannot help but wonder about the reasoning behind their odd shapes: e.g. 'Winston's Hiccup' on the border between Jordan and Saudi Arabia.

'Natural' and 'artificial' borders

So-called natural borders represent around 55% of the global total: borders in the middle of a river or a stretch of water (hydrographic features, 30%) or a mountain chain or valley (orographic features, 25%). These often follow the crest line, the watershed line or the thalweg (the line connecting the lowest points in a river bed or valley). For a long time, deserts, forests and marshes were seen as possible natural borders as they were difficult to cross; now, however, that is no longer the case, and these features also have the disadvantage of being stretches of land, rather than lines.

So-called artificial borders, 45% of the total, are often straight lines (25%), as is the case for many borders in North America and Africa. They often follow lines of longitude (Egypt–Sudan) or latitude (Alaska, Korea).

But can any border be called natural? Some even try to call these borders 'scientific', implying that they are both identifiable and defendable. This idea is very much open to question. Mountains and rivers make easily defendable borders, but why should that be an essential criterion in the 21st century? And when a mountain range separates two countries, does the natural border run along the crest line or the watershed line? Even in the latter case, it's not always easy to determine (Thailand–Cambodia). Rivers and lakes are easily identifiable boundaries, which can help to avoid disputes. But rivers bring communities together more than they divide them. And where should the border be drawn: on the bank, on the median line, along the thalweg? What's more, who should its islands belong to? China and Russia took forty years to come to an agreement over ownership of 2,444 river islands! And then there is the fact that rivers can change course. When that happened to the Rio Grande, it sparked a border dispute between the US and Mexico that lasted a century (until 1967), and the alluvial plain has been open since 1970 to avoid any new disputes. Austria and Italy adopted the concept of the 'moving border' to take into account changes to the watershed line due to climate change. Changes to the course of the Danube are at the heart of an ongoing border dispute between Croatia and Serbia.

When it comes down to it, it is not always possible to draw a 'natural' border. There are not always rivers, lakes or mountains in convenient locations. What about the US and Canada, the Arabian Peninsula or Russia's southern border? So-called natural borders are by no means less complicated, arbitrary, unjust or disputed. The mountain border between Afghanistan and Pakistan divides the Pashtun people; the Pyrenees divide the Basque community as well as the Catalans. Defining Europe's eastern border as the Urals doesn't reflect any historical or human reality. Neither is there a reason to consider the Bosporus Strait or the Dardanelles as the southernmost edge of Europe – especially because Türkiye straddles them. The Rhine, the Oder and the Neisse are among the most disputed borders in European history. Trying to divide people up within these borders in a coherent, uncontroversial way would mean ignoring the realities of human geography.

On the other hand, so-called 'artificial borders' can be simple, legitimate, fair and peaceful. The straight lines drawn through the middle of the Sahara don't create much tension, as long as they can be easily crossed. For many years, the Tuareg people treated the Sahara as effectively borderless – a kind of sandy Schengen Area. And colonial borders – in Africa, the Middle East, Asia – did not always ignore political and ethnic realities – far from it, in fact. Iraq is made up of former vilayets (provinces) of the Ottoman Empire. Only 700 km of borders in the region were drawn up by the Sykes–Picot Agreement in 1916. The Iraq–Iran border dates back to 1639; the border between Iran and Türkiye, to 1514. Discourse about the supposed 'artificiality of African borders' is a colonial concern: it is not pushed by those in power on the continent itself. What's more, former colonies also draw up completely 'artificial' borders, designed to protect economic interests and separate communities (Morocco–Western Sahara). It should also be noted that it is not possible to draw a line that neatly and uncontroversially separates a particular people, nation, ethnic or linguistic group from others around it. Every border has an element of arbitrariness.

The concept of 'natural borders' also has another meaning: a deliberate political project, an ideal to aim for. France, bordered by mountains and rivers, has always been sympathetic to this idea: Julius Caesar was the first to speak of the 'natural' borders of Gaul. As an ideology, it can be traced back to Richelieu. Danton told the French National Convention: 'The limits of France are marked by nature, we will reach the four corners of the horizon, to the edge of the Rhine, to the edge of the ocean, to the edge of the Pyrenees, to the edge of the Alps'. Later, this ideology also relied on the concept of a community with a shared language (e.g. Alsace-Lorraine). The idea of natural political borders has fuelled expansionism and imperialism – the borders drawn up were often those that allowed the territory to be most easily defended – as much as

So-called 'artificial borders' can be simple, legitimate, fair and peaceful.

decolonization (Woodrow Wilson's Fourteen Points).

'In summary, all borders are artificial, in the sense that they are defined by men and are therefore arbitrary: they are scars left by history' (François Terré). 'Nature is absolutely innocent of the borders that we accuse it of having created' (Pierre Larousse). 'All political boundaries are man-made, that is, artificial; obviously, they are not phenomena of nature' (Richard Hartshorne). 'Mountains and waterways have a suggestive power, but they can only be elevated to the status of borders through a solemn act of writing, the only way to transcribe an accident of nature into law' (Régis Debray). 'There is no such thing as a good border in absolute terms, let alone an ideal border, only real borders that are either mutually recognized as legitimate or provide more political, strategic and economic advantages for one side than the other, at a given moment in history' (Michel Foucher).

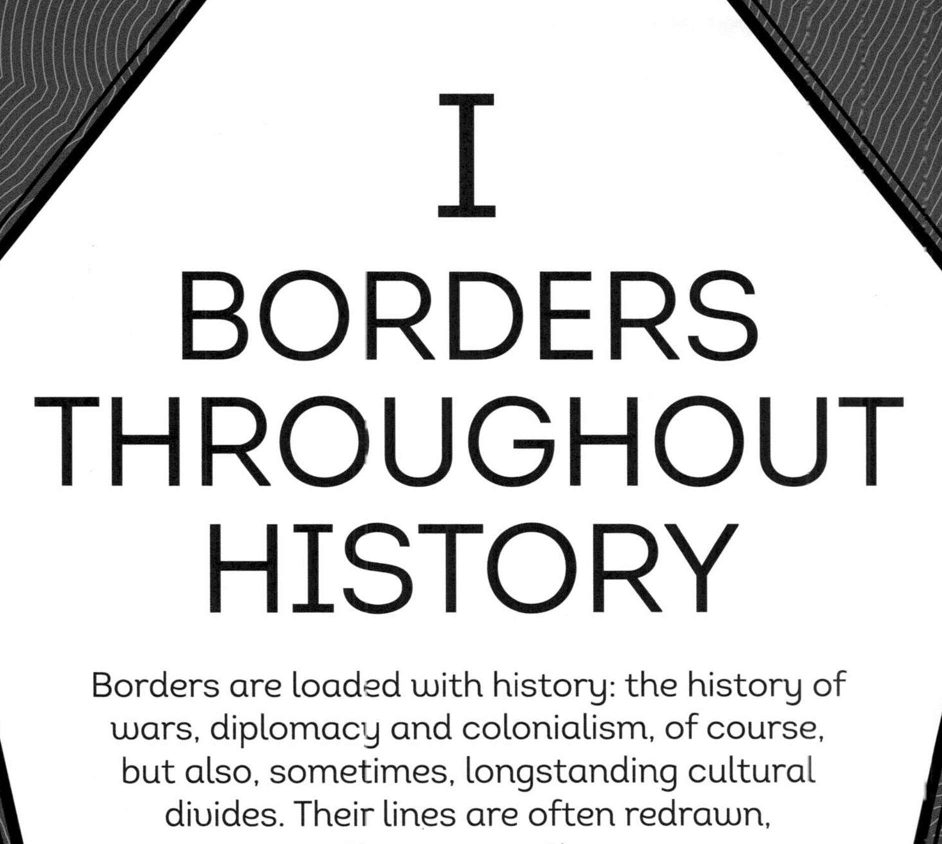

I

BORDERS THROUGHOUT HISTORY

Borders are loaded with history: the history of wars, diplomacy and colonialism, of course, but also, sometimes, longstanding cultural divides. Their lines are often redrawn, sometimes many times as the years go by.

I · BORDERS THROUGHOUT HISTORY

In ancient times, borders were often not clear lines, but vaguely defined areas. The outer limits of empires were fringes or frontier districts known as 'marches', a term that originally referred to fiefs in the Carolingian Empire. Markers such as posts, stelae or boundary stones represented dominion but did not serve as dividing lines. Walls built by the Romans and Chinese served as fortifications or barriers used to control trade and taxes, but did not necessarily mark a line between 'them' and 'us', or 'barbarians' and the 'civilized'; indeed, they sometimes divided communities. It was also possible for control over territory to shift in stages, rather than abruptly. In the ancient Chinese worldview, the ruler or emperor stood at the centre, surrounded by royal domains, which were in turn surrounded by the domains of lords and princes, partially civilized borderlands, 'allied' barbarians, and in the outermost circle, 'savages'.

The Roman Limes (a term originally used by land surveyors, meaning 'border path') illustrated a similar ancient understanding of borders, imposed unilaterally. They were not unbroken lines, nor were they the same everywhere. In the north of Europe, the Limes line was closed by Hadrian's Wall and the Antonine Wall, marking the end of the conquered land. Between the Rhine and the Danube, the best-known Limes was made up of a network of parallel and perpendicular roads with clay ramparts. In Africa, the Roman Limes was represented by the Fossatum Africae or 'African ditch', a series of ditches and embankments constructed along the desert's edge, made up of several sections.

The idea of building a border around a territory is not a given. In the ancient world, most countries were sparsely populated, and mountains often formed a natural obstacle. In Islamic tradition, the outer border of Dar al-Islam was temporary, and state borders within it were unlawful. In Chinese tradition, the only legitimate borders were those imposed by China itself.

The emergence of modern borders

The emergence of borders as we know them today is inextricably linked to the birth of the modern world. Their roots go back to the 17th century. The Peace of Westphalia (1648) gave rise to the first borders negotiated between empires, the Treaty of the Pyrenees (1659) being the first example. The modern delimitation of borders between states was born from negotiations about the border between Russia and Türkiye. Borders developed along with modern cartography: the more detailed the map, the more accurately boundaries can be drawn. At that time, while borders could be used to monitor and restrict entry – especially when dealing with infectious disease – they were primarily intended to monitor people leaving national territory, and stop those fleeing conscription or taxes. But in the time of empires, the idea of a border also still referred to an area yet to be conquered (a frontier) or a 'buffer' zone, such as the neutral zone established by Great Britain and Germany in the

Borders developed along with modern cartography: the more detailed the map, the more accurately boundaries can be drawn.

Gold Coast in 1888, or the demilitarized zone set up on the bank of the Mekong during the Franco–Siamese War in 1893. Geographer Friedrich Ratzel wrote: 'The borderlands are the reality, the boundary line is an abstraction thereof.'

There are very few borders that date to before 1800: these older borders are mostly in Europe, although the Amur River (Russia–China) is also example. In some cases, modern borders follow the line of ancient ones, for example in Latin America and South East Asia.

Between the mid-19th century and the eve of the First World War, the whole world was divided up, in parallel with the establishment of nation-states: in terms of length, more than half of all land borders date to the years between 1875 and 1924. Around the same time, the United States (1890) and Russia (1860) completed the 'conquest' of their huge respective territories. Borders served their traditional functions: defending territory, levying taxes, imposing customs duties, ensuring public order. There were 24 nation-states in 1914, and 48 members of the League of Nations in 1920; the UN had 51 members in 1945, and it now has 193. Since 1945, the number of borders has increased dramatically, due to the end of empires and the fall of the Eastern bloc. Almost 30,000 km of borders have been drawn since 1989 (including 48 new dyad borders, which are those between two countries). More than 10% of current borders were created after 1990.

Virtual borders

Cultural borders (between languages, religions, 'civilizations') rarely align with state borders, but they may exist in people's minds and hearts. Legal borders are not always seen as legitimate.

Between 1945 and 1990, as well as the East–West divide, the idea of a North–South divide between developed and developing countries was often alluded to, although it was difficult to draw a clear geographical line of separation. After the Cold War, it was tempting to adopt a new way of dividing up the world, but the idea of a rich 'West' shutting itself off from a poorer 'South' is too much of a generalization to be useful.

Can we trace dividing lines between large cultural areas? The concepts of 'East' and 'West' have never been more than conventional distinctions (e.g. the Roman Empire, Christian churches), although the delimitation of religious areas of influence is sometimes a legacy of imperial borders (e.g. the Rhine, the Danube). Where does Asia begin? The Urals and the Bosporus Strait are arbitrary limits; the river Indus has a more convincing claim. The idea of

Evolution of the Germany–Poland border

the 'Near East' is French in origin; the British notion of the 'Middle East' is broader than the Mashriq ('the East'), which refers to the countries in the Fertile Crescent and represents the centre of Arab civilization, as opposed to the more distant Maghreb (meaning 'sunset'). Then there is the 'Greater Middle East', a term introduced by the US, which referred to a vaguely defined stretch of land from the Atlantic to India.

Cyber borders are, by their very nature, virtual: while it is difficult to completely cut off a country's access to the internet, it can be restricted by controlling access to the fibreoptic cables that connect it to the outside world, which are often limited in number: China is one such example.

Contemporary borders

Today there are 315 international land borders separating two states (dyads), stretching for a total of 261,500 km. The most recent borders to be recognized are those of South Sudan. If we add

Two fundamental principles of modern international law are the inviolability of borders and territorial integrity.

in maritime borders, whether they are clearly delimited or not, the global total comes to around 750 borders between states. Most countries share borders with a number of others, and countries that only border one other country are relatively rare: Ireland (with the UK), Lesotho (with South Africa), Monaco (with France), San Marino (with Italy), the Vatican City (with Italy).

There are very few new land borders being drawn up today. Why is that? Firstly, because there are very few unclaimed zones left on the globe: 'The time of the finished world has begun,' in the words of Paul Valéry. Secondly, because norms of behaviour have been established. Borders are mainly drawn up by agreement between the interested parties, rather than unilaterally imposed. Territories are no longer being carved up by force: since 1945, only 30% of territorial wars have resulted in new territorial divisions, as opposed to 80% before that date. The border between Germany and Poland, which has been redrawn a number of times since 1939, is now stable.

The concept of irredentism still exists – even in Europe, with Hungary and its former empire – but it is rarely put into practice now: since the mid-1970s, there have been very few instances of territory being annexed by force. There were 14 examples in the 1980s, 10 in the 1990s, only 2 in the 2000s and again in the 2010s (the occupation of the Costa Rican island of Isla Calero in 2010, the annexation of Crimea in 2014). Those who try to conquer territory by force run the risk of provoking a military response (the Falklands, Kuwait) or political stigmatization (the Caucasus, Crimea, Donbas). Instead, governments sometimes issue passports in occupied zones (Russia in the Caucasus and in Ukraine). There are around a dozen territories considered to be under occupation by at least one state.

Two fundamental principles of modern international law are the inviolability of borders and territorial integrity. Although these are often confused, the first means it is illegal to cross a border without permission, while the second means it is illegal to attack the territory of another nation-state.

A third fundamental principle is the intangibility of borders, which is more a custom than a law. Summed up by the Latin expression *uti possidetis, ita possideatis* ('as you possess, so you may possess'), it dates back to the era of decolonization, and more specifically, when Latin American countries gained independence in the early 19th century. It means that, even if they are unfair and imperfect, maintaining existing borders is preferable to completely redrawing a border. 'They should only be touched with a trembling hand,' in the words of Montesquieu.

This principle – which should not be taken to mean that borders are immutable and can never be changed – has been inscribed in the Charter of the Organization of African Unity (although it was rejected by Somalia and accepted with reservations by Morocco), as well as various international agreements. It is the reason why the International Court of Justice has generally confirmed imperial borders when ruling on disputes that have been brought before it. And it is why all borders drawn up since 1989 – whether by separation, independence or secession – have followed existing lines (administrative divisions). This is not always without its problems (for example, the Serbian minority in Kosovo). But it is the least controversial choice – and the one least likely to spark a conflict. However, when these provincial divisions have not been clearly delimited, disputes can arise. That is one of the reasons why two breaches of the principle of the intangibility of borders in Africa have led to bloody conflicts (Ethiopia/Eritrea, Sudan/South Sudan).

It should also be noted that questions around borders are an exception when it comes to treaty law. A fundamental principle of treaty law, governed the Vienna Convention (1969), is *rebus sic stantibus* ('things standing thus'): a treaty, like any contract, is only valid if the circumstances that were in place when it was signed remain. However, Article 62.2 of the convention says: 'A fundamental change of circumstances [...] may not be invoked as a ground for terminating or withdrawing from a treaty [...] if the treaty establishes a boundary.' Ultimately, it could be said that 'the border creates itself': as time goes by, existing borders become more widely accepted.

It is clear that we are in an era of consolidation when it comes to borders. They are generally delimited (i.e. the overall line of the border is agreed) with reference to topography; they are increasingly demarcated (given physical form through boundary markers and barriers) with the help of precise geo-coordinates (using GPS); and they are defended (surveillance technology, barriers and walls). 'Never has there been so much negotiation, delimitation, demarcation, definition, equipping, monitoring and patrolling of borders' (Michel Foucher). It is true that the borders inherited from decolonization, in particular, were often vague. Since the turn of the century, therefore, a number of border disputes have been resolved and borders drawn up, sometimes through arbitration (the International Court of Justice – in Africa especially; the Permanent Court of Arbitration; ad hoc tribunals). In principle, border settlements have to be declared to the UN, according to Article 102 of the Charter.

In fact, borders are often adjusted. In 2007, a narrow strip of land known as Abrene in Latvian and Pytalovo in Russian was legally signed over to Russia, which had occupied it since 1991. In 2016, Egypt ceded the islands of Tiran and Sanafir to Saudi Arabia. In the same year, Belgium and the Netherlands signed an agreement to swap pieces of territory along the river Meuse.

Geographers, legal experts and surveyors will not be finding themselves out of work any time soon. In Africa especially, the continent with the greatest total length of borders (83,500 km), almost two thirds have either not yet been delimited (one third) or demarcated (one third; the African Union is working on an ongoing border demarcation project). This focus on borders can be explained by increased movement (migration, trade) and risks (trafficking, terrorism, epidemics, tax evasion), but also by the modernization of states and technological advances: observation and surveillance technology (satellites, drones, CCTV), measuring instruments (the use of lasers to draw boundaries) and simulations (computer modelling), which all make it easier to monitor borders. What's more, international organizations are reluctant to accept member states with unresolved border issues. The demarcation of borders is not always simple: it may conflict with the way of life of nomadic communities, or be complicated by the natural movement of rivers, for instance.

Almost all land on earth now falls under the sovereignty of a state, and extraterritorial zones are very rare. Contrary to popular belief, diplomatic embassies are not truly extraterritorial: they are part of the national territory on which they stand. The same is true of free economic zones and the international zones in airports (including customs, transit and waiting areas). When it comes to planes and ships, the situation is a little more complicated: they are subject to the law of the country whose national waters or air space they are passing through (with certain privileges), but subject to the law of their country of registration when in international waters or air space.

Borders divide, but they can also unite. The movement of goods and citizens creates places where people come together. There are a dozen or so 'twin cities' along the border between the US and Mexico, forming an area with 13 million inhabitants known as the Tercera Nación ('Third Nation') or Mexamérica. On the Mexican side, it is home to thousands of *maquiladoras*, delocalized US companies. In Africa, there is southern Senegambia, the SKBo cross-border area (Mali, Côte d'Ivoire, Burkina Faso) and the Niger–Nigeria border. Then there are the 'peace parks', cross-border nature reserves created by South Africa and its neighbours

Ultimately, it could be said that 'the border creates itself': as time goes by, existing borders become more widely accepted.

– examples of which also exist in Europe (Poland–Czechia) and North America (Canada–United States). In Europe, there are many 'spaces of cross-border cooperation'. Sometimes these create bizarre situations: in Baarle, there is a shop that straddles the border between Belgium and the Netherlands and is therefore sometimes half-open, depending on local opening hours. Even states that have tensions between them can be pragmatic on a local level. Road 178 in Estonia briefly crosses a piece of Russian territory called the Saatse Boot: vehicles are allowed to pass through freely as long as no one gets out.

Today, borders are often an area as much as a line. Border regions are places of exchange, of trade and trafficking,

> There is a possibility that new borders could be drawn within the European Union, if a region such as Catalonia were to declare independence.

and temporary migration, sometimes on a daily basis (cross-border workers). Borders are places of surveillance and monitoring (a 'thick' border), which can take the form of a no-man's-land or an area with a particular customs status. The desire to create a buffer zone, in the traditional sense of the word, was part of Israel's logic in South Lebanon (1982–2000), as well as Russia's military presence outside its national territory (Transnistria, the Caucasus, Ukraine), China's attachment to Tibet and Pakistan's to Afghanistan. The idea of a buffer zone was made real in Pakistan, which inherited from the British colonizers a 'province on the north-west border' (later renamed Khyber Pakhtunkhwa) and, on the Afghan border itself, 'tribal areas under federal administration' (these were brought under provincial administration in 2018). Neighbouring states of an unstable country often try, at least temporarily, to create buffer zones on their territory, as Algeria and Tunisia have done for their borders with Libya.

While many borders are subject to ever greater monitoring and surveillance, there are very few that are completely closed; Morocco–Algeria since 1994, North Korea–South Korea. And while some borders are breached, crossed or disputed, the majority remain peaceful.

The borders of Europe

The European continent has 100 borders, stretching a total length of 37,000 km. They include some of the oldest borders on the planet: Andorra (the oldest in the world, created on 8 September 1278), San Marino, Switzerland, Spain–Portugal, France–Spain, Norway–Sweden, Germany–Czechia. Others date to the end of the First World War, while after the Second World War, the preferred approach was to displace people rather than redraw borders. Others are more recent and date to the collapse of the Eastern bloc. There is a possibility that new borders could be drawn within the European Union, if a region such as Catalonia were to declare independence, or a country was split in two (e.g. Flanders and Wallonia in Belgium) – and, of course, that's without taking future EU expansion into account.

The European Union is the place where two major border issues come together: that of removing internal borders (the Schengen Area) and protecting external ones (the Frontex agency and the European border surveillance system Eurosur).

There are very few border disputes remaining within Western Europe, and those that exist are very minor. The longstanding dispute between the UK and Ireland over the island of Rockall was resolved in 2014. The absence of a recognized border between Germany, Austria and Switzerland in Lake Constance is not a source of tension between the countries concerned. The question of South Tyrol has not troubled relations between Italy and Austria since the dispute was officially resolved in 1992. Spain and Portugal enjoy good relations despite their dispute over the region of Olivenza. Slovenia and Croatia are trying to come to an agreement over the delimitation of their border in the Gulf of Piran. The European Union even requires all future members to resolve their border disputes before joining, although that rule has not been enforced in the case of Cyprus. When it comes to its five enclave states (Andorra, Liechtenstein, San Marino, Switzerland, Vatican City), there are no significant territorial disputes. Minor border adjustments are common – for example, an exchange of territories between Belgium and the Netherlands took place in 2016.

Outside of Western Europe, however, the question of borders was reopened by Russia's occupation of Ukrainian territories. Meanwhile, Serbia and Kosovo are planning a significant exchange of land, which could lead to other territorial claims in the Balkans.

The question of Europe's external borders, however, is particularly complex and problematic. We should distinguish between the borders of the Schengen Area, those of the European Union (EU) and those of Europe as an idea or concept. In 2024, the Schengen Area included 25 EU countries and 4 non-member states. The borders of the European Union have changed over time, as it has expanded. And the boundaries

> The question of Russia sums up all the ambiguities inherent in the idea of Europe. Russia itself has always been uncertain: is it European or Eurasian?

of the EU are not clearly defined. Firstly, because there are areas where they are disputed: Morocco claims the Spanish enclaves and Perejil Island. Greece and Türkiye have not delimited their maritime border. Most significantly, Cyprus has joined the EU when a large part of its territory is still occupied.

What's more, what exactly do we mean by the borders of Europe? Do we stop at the land or maritime borders of the main territories of member states: the Italian island of Lampedusa, the outer limit of Cyprus's exclusive economic zone? But that wouldn't take into account the outermost regions of the EU: the French overseas departments, the islands that make up the autonomous regions of Portugal, the autonomous community of the Canaries. On the other hand, the British military bases in Cyprus and the British Crown Dependencies were not considered part of the EU when the UK was a member. In all, the EU has 14,000 km of land borders.

How far could the borders of the EU reach in the future? There is no definition of what constitutes a European country laid out in the treaties. While the UK has left the EU, Norway, Iceland and Switzerland, all members of the European Free Trade Association (EFTA), would surely be accepted if they wanted to become members. From Brussels's perspective, there is little doubt that all the states of the former Yugoslavia may join the EU one day. These countries would, however, need to renounce any claim to territory they do not currently hold. Would the EU allow Kosovo and Albania to unite, at the risk of destabilizing the former Yugoslavia? Could it, in order to resolve the problem of the Serbian minority in northern Kosovo, approve an exchange of territory with Serbia (the Preševo valley, where the population is mainly Albanian), at the risk of reopening the 'Pandora's box' of borders in the Balkans?

Beyond the Balkans, the question of the limits of Europe is a never-ending debate. Here is where the concept of Europe needs to be interrogated.

To the south, things are relatively simple: it is unlikely that the countries of North Africa would ever claim a European identity. To the east, the situation is a little more complex, but the countries of the South Caucasus (Armenia, Georgia) are generally considered part of Europe from a historical and cultural perspective. Other former Soviet republics (Belarus, Ukraine, Moldova) also have close ties to the continent's history and culture.

It is in the south-east that things start to get really complicated. With all due respect to the tourists who like to take pictures of themselves on the bridges of Istanbul with 'one foot in Europe and one foot in Asia', it is too simplistic to say that Europe ends at the Bosporus Strait. Firstly because, if we were to accept this definition, part of Türkiye (Eastern Thrace) would indisputably lie within Europe. Secondly because the history of Türkiye is inextricably linked to that of the rest of the continent. For all that, could modern Turkish culture tie the entire state to Europe?

The question of Russia sums up all the ambiguities inherent in the idea of Europe. Russia itself has always been uncertain: is it European or Eurasian? 'Russia doesn't really know where it begins or where it ends,' said Václav Havel. Which is why, throughout history, there have been various attempts to define the borders of 'European Russia' in different places. In the 16th century, it was the Don river. Then the boundary moved to the Gulf of Ob, in the north-east. In the 18th century, Russia's imperial expansion made this definition problematic, but geographers were reluctant to treat the entire empire as European. In 1730, a Swedish geographer, Philip Johan von Strahlenberg, proposed an arbitrary compromise for the dividing line: the Urals. In Russia, and across the rest of the continent, this definition was accepted: the Urals remain the de facto border of Europe. Even today, some Russian roads are seen to represent 'crossing from Europe to Asia'.

There is no single possible definition of the borders of Europe. Geologically speaking, Europe is not a continent: there is a Eurasian Plate, but this doesn't include Cyprus, Malta, Sicily or half of Iceland. Culturally, Europe is a melting pot without clearly defined boundaries. The Croatians see themselves as the last bastion protecting Europe against the Serbs. The Serbs, in turn, sometimes consider themselves the last defence against the Albanians and the Turks; a view that the Greeks are also inclined towards. But many Turks consider themselves European. Historically, the word 'Europe' was first used as a name for the region of Southern Thrace, then with the advance of the Turks and Tatars, became a term for the Western Roman Empire. Politically, Europe at very least consists of the European Union and the four states of the European Free Trade Association, plus the UK – but not all the members of the Council of Europe, which included Russia up until 2022, and still includes Türkiye and the states of the Caucasus region.

Sources: *Atlas des civilisations*, Paris: Le Monde/La Vie, 2015; Samuel P. Huntington, *The Clash of Civilizations and the Remaking of World Order*, New York: Simon & Schuster, 1996; Jean-Christophe Rufin, *L'Empire et les nouveaux barbares*, Paris: JC Lattès, 2001.

Black Sea
RUSSIAN EMPIRE
Caspian Sea
Constantinople
Trebizond
Ankara
Baku
ANATOLIA
Tabriz
Adana
Tigris
SYRIA
Aleppo
Mosul
Kirkuk
Euphrates
PERSIA
Beirut
Mediterranean Sea
Damascus
Baghdad
Acre
IRAQ
Jerusalem
Amman
Alexandria
Cairo
Kuwait
Nile
Persian Gulf
EGYPT
Luxor
ARABIA
Dubai
Riyad
Medina
Muscat
Mecca
ANGLO-EGYPTIAN SUDAN
Red Sea
Indian Ocean
Khartoum
500 km

THE SYKES-PICOT AGREEMENT

A sharing of spheres of influence

The historical Sykes-Picot 'line' made the news in the 21st century when ISIS 'opened' the border between Syria and Iraq in 2014; as a symbolic act, an Islamic State fighter was filmed removing its physical boundary markers. But this reference to the reviled colonizers is not entirely accurate. In 1916, Sir Mark Sykes and François Georges-Picot did not draw a border, but instread agreed on Britain and France's respective spheres of influence, from the Mediterranean to the Euphrates, with a dividing line that ran 'from the "e" in Acre to the final "k" in Kirkuk'. Sykes, ever cautious, signed the agreement in pencil. The accord was not respected everywhere and most borders in the Middle East were established by the San Remo Conference (1920) and the Treaties of Sèvres (1920) and Lausanne (1923).

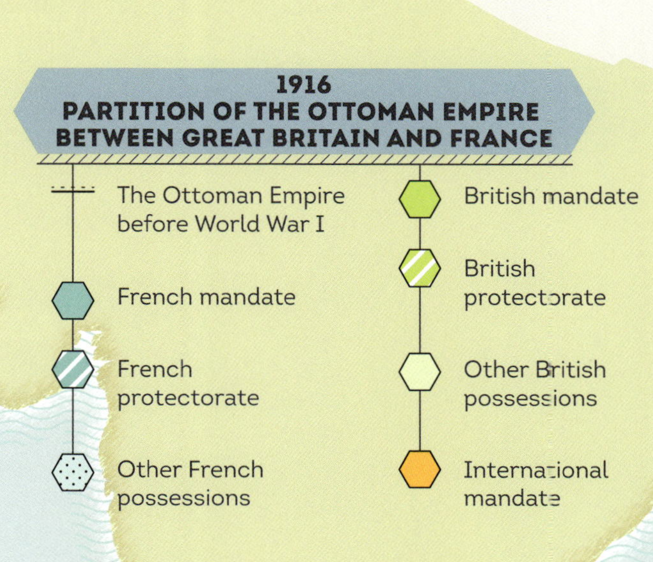

Islamic State's 2014 attempt to destroy the current border between Iraq and Syria

Source: 'Map of Eastern Turkey in Asia, Syria and Western Persia', Royal Geographical Society, 1916.

5 Macartney-MacDonald Line
CHINA
KHYBER
PAKHTUNKHWA
GILGIT-
BALTISTAN
AKSAI
CHIN
2 Johnson Line
AFGHANISTAN
NORTH
KHORASAN
AZAD
KASHMIR
JAMMU
AND KASHMIR
TIBET
4 Durand Line
PUNJAB
6 MacMahon Lin
IRAN
7 Radcliffe Line
NEPAL
BHUTAN
ARUNACHAL
PRADESH
3 Goldsmid Line
BALUCHISTAN
RAJASTHAN
SOUTH
KHORASAN
PAKISTAN
7 Radcliffe Line
BANGL.
SIND
BENGAL
GUJARAT
Gulf of
Oman
INDIA
1 Pemberton Line
MYANMA
(BURMA
Gulf of
Bengal
INDIAN
OCEAN

THE 'LINES' OF SOUTH ASIA

A legacy of the British Empire

The borders drawn up by the British colonizers in South Asia are some of the most contentious on the planet. Strained relations between the states in the region have not helped to legitimize these borders, which are often of recent origin and generally disputed. The Goldsmid Line (Iran–Pakistan, 1872, later modified) today makes up part of the border between Iran and Pakistan. The border between Afghanistan and Pakistan more or less follows the Durand Line (1893), which was intended to separate conflicting tribal communities. But Kabul has disputed its validity since 1949 and the two parties have never agreed on the exact line of their shared border. The India–Pakistan border follows the Radcliffe Line (1947) from Gujarat/Sindh to the Punjab. The McMahon Line (1914), which separates Tibet from China to the east, is not recognized by Beijing, which also disputes Indian sovereignty over Arunachal Pradesh, while China's occupation of Aksai Chin is disputed by India, which recognizes the Macartney–MacDonald Line (which superseded the Johnson Line) as the valid border.

1. **1837: Robert Boileau Pemberton**, *British army officer*
2. **1865: William Johnson**, *British topographer*
3. **1871: Sir Frederic Goldsmid**, *British army officer*
4. **1893: Sir Mortimer Durand**, *British diplomat*
5. **1899: Sir George Macartney**, *British governor and diplomat* *and* **Sir Claude Maxwell MacDonald**, *British diplomat*
6. **1914: Sir Henry MacMahon**, *British officer and diplomat*
7. **1947: Sir Cyril Radcliffe**, *British lawyer*

Sources: J. G. Bartholomew, 'South-Western Asia,' *The Times*, 1922; Nigel Dalziel, *The Penguin Historical Atlas of the British Empire*, London: Penguin, 2012.

THE CURTAINS OF THE COLD WAR

Political divisions across continents

CURTAINS OF DIVISION

THE CACTUS CURTAIN
Cuba

THE IRON CURTAIN
Europe

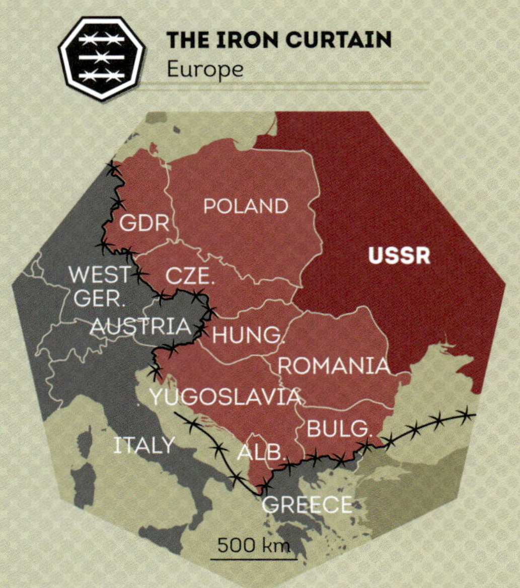

THE ICE CURTAIN
Arctic

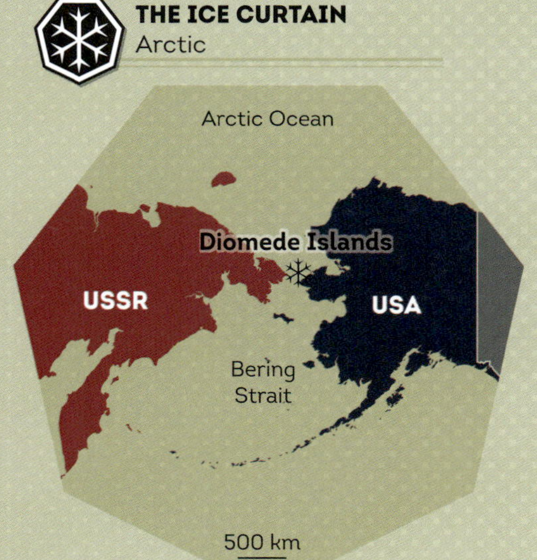

THE BAMBOO CURTAIN
Asia

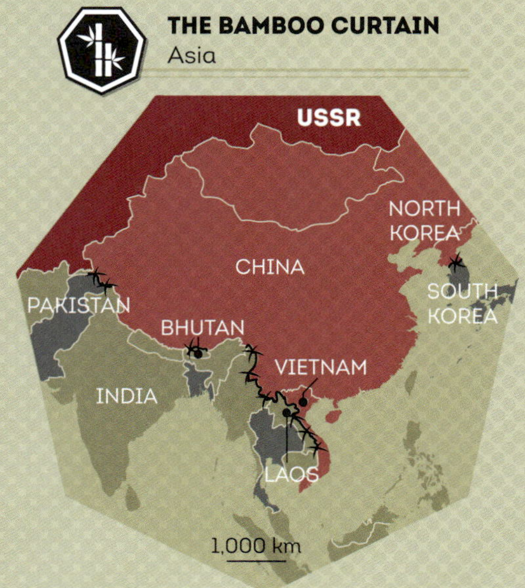

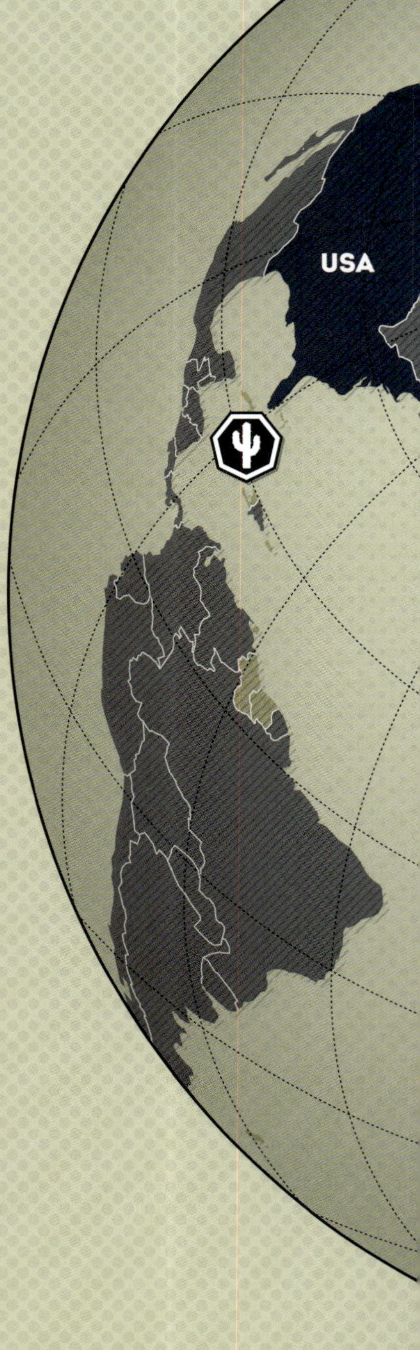

THE USA AND ITS ALLIES
Western Bloc

A POLITICAL BORDER
1990
Western countries
Communist Bloc
ICELAND
Reykjavik
NORWAY
Oslo
SWEDEN
Stockholm
FINLAND
Helsinki
North Sea
DENMARK
Copenhagen
UNITED KINGDOM
London
IRELAND
Dublin
NETHERLANDS
Amsterdam
Brussels
BELGIUM
Bonn
LUX.
Berlin
EAST GERMANY
POLAND
Warsaw
Prague
CZECHOSLOVAKIA
WEST GERMANY
Vienna
AUSTRIA
Budapest
HUNGARY
ROMANIA
Bucharest
Moscow
USSR
Black Sea
Paris
FRANCE
Bern
SWITZERLAND
Atlantic Ocean
Belgrade
YUGOSLAVIA
BULGARIA
Sofia
ALBANIA
Tirana
ITALY
Rome
SPAIN
Madrid
PORTUGAL
Lisbon
GREECE
Athens
Ankara
TURKEY
Nicosia
CYPRUS
SYRIA
LEBANON
IRAQ
Mediterranean Sea
MOROCCO
ALGERIA
TUNISIA
LIBYA
EGYPT
500 km
Sources: Georges Duby, Grand Atlas historique, Paris: Larousse, 2011; Michel Foucher, Fragments d'Europe, Paris: Fayard, 1993.

THE BOUNDARIES OF EUROPE

Multiple points of view

The borders of the European Union have changed over time, as some countries have joined and others have left. Its political boundaries remain less clear, however, because in some places they are disputed (the Aegean Sea, Cyprus). But what about the geographical boundaries of Europe itself? In the south, the Mediterranean makes things relatively simple. But in the south-east, the situation becomes more complicated. It is hard to say that 'Europe ends at the Bosporus Strait', when it runs right through the centre of Türkiye. In the east, the former Soviet republics are historically and culturally part of the continent of Europe. But the question of Russia sums up all the ambiguities inherent in the idea of Europe. Russia itself has always been uncertain: is it European or Eurasian? In 1730, a Swedish geographer proposed an arbitrary compromise: the Urals.

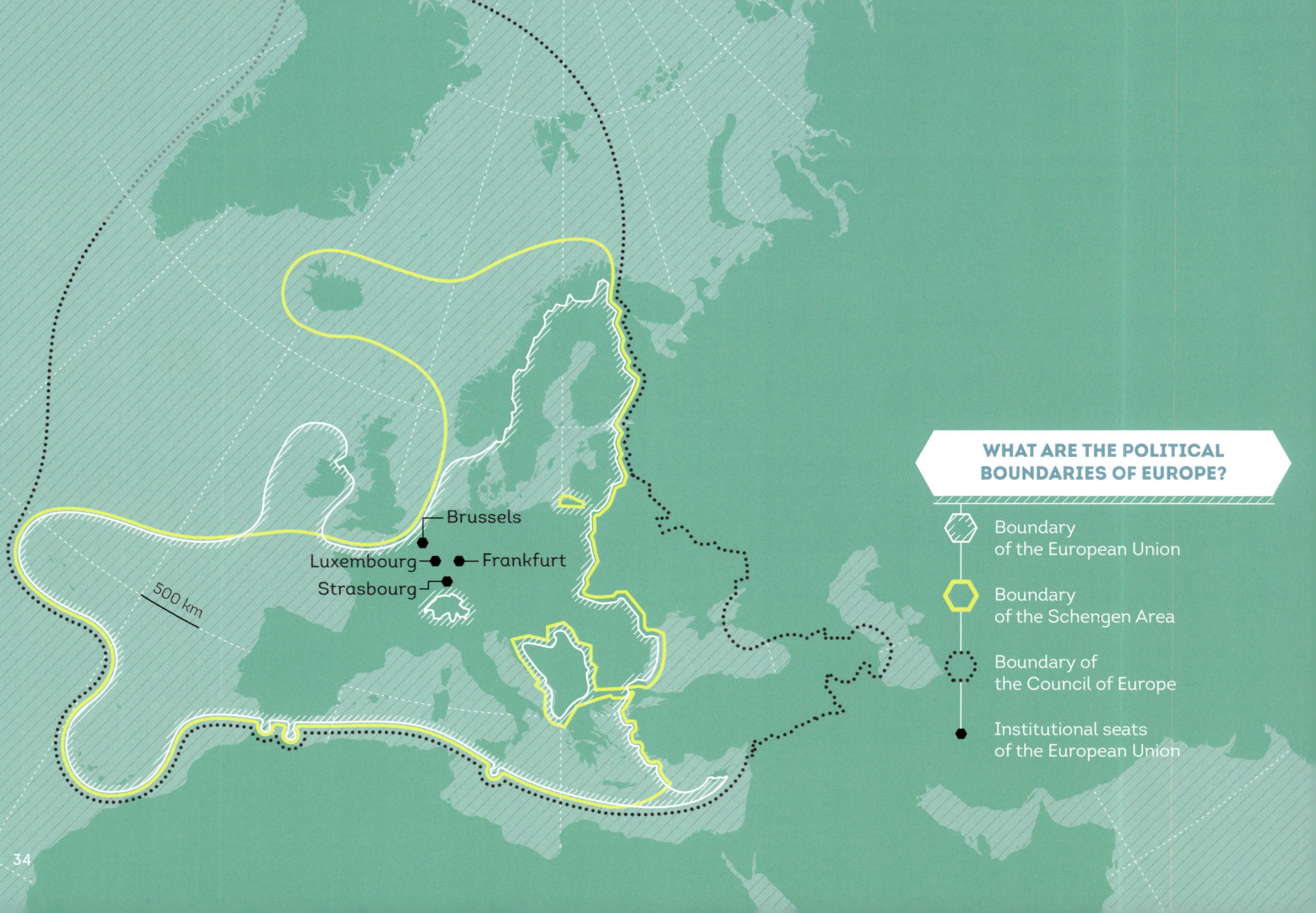

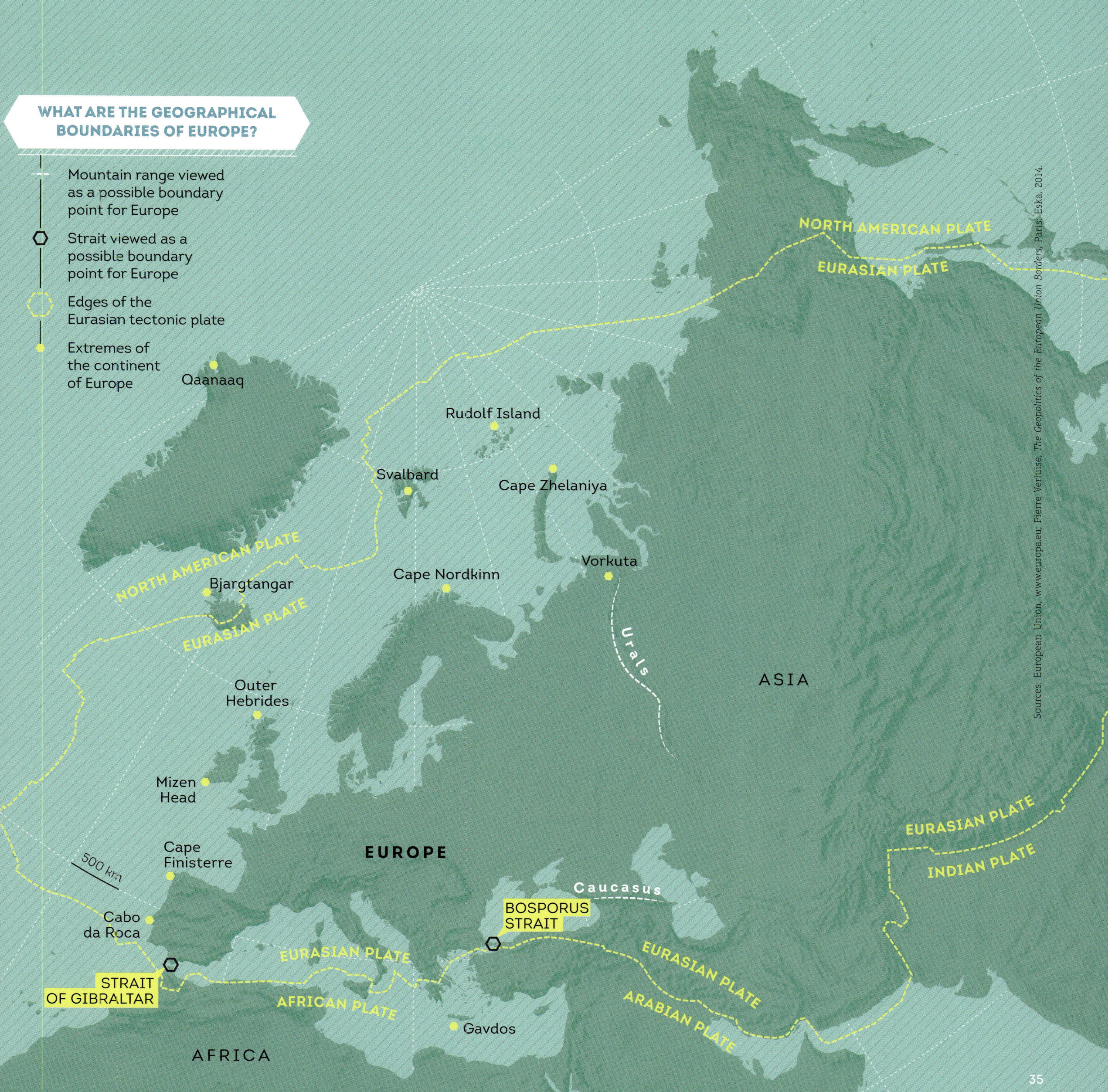

Sources: European Union, www.europa.eu; Pierre Verluise, *The Geopolitics of the European Union Borders*, Paris: Eska, 2014.

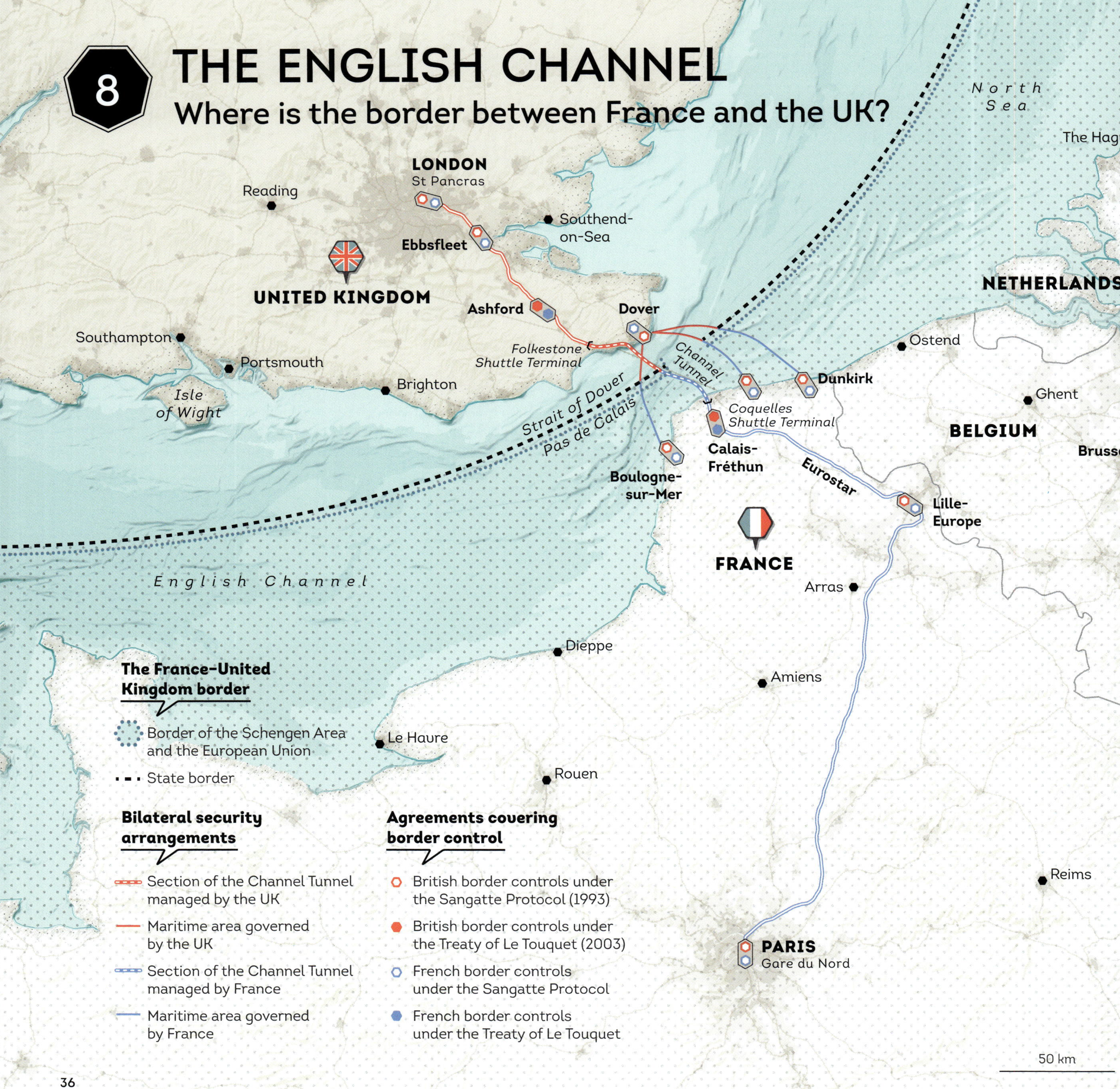

8
THE ENGLISH CHANNEL
Where is the border between France and the UK?
North Sea
The Hagu
LONDON
St Pancras
Reading
Southend-on-Sea
Ebbsfleet
UNITED KINGDOM
NETHERLANDS
Ashford
Dover
Southampton
Portsmouth
Folkestone Shuttle Terminal
Channel Tunnel
Ostend
Dunkirk
Isle of Wight
Brighton
Ghent
Coquelles Shuttle Terminal
BELGIUM
Strait of Dover
Pas de Calais
Calais-Fréthun
Brusse
Boulogne-sur-Mer
Eurostar
Lille-Europe
FRANCE
English Channel
Arras
Dieppe
Amiens
The France–United Kingdom border
Border of the Schengen Area and the European Union
State border
Le Havre
Rouen
Bilateral security arrangements
Section of the Channel Tunnel managed by the UK
Maritime area governed by the UK
Section of the Channel Tunnel managed by France
Maritime area governed by France
Agreements covering border control
British border controls under the Sangatte Protocol (1993)
British border controls under the Treaty of Le Touquet (2003)
French border controls under the Sangatte Protocol
French border controls under the Treaty of Le Touquet
Reims
PARIS
Gare du Nord
50 km

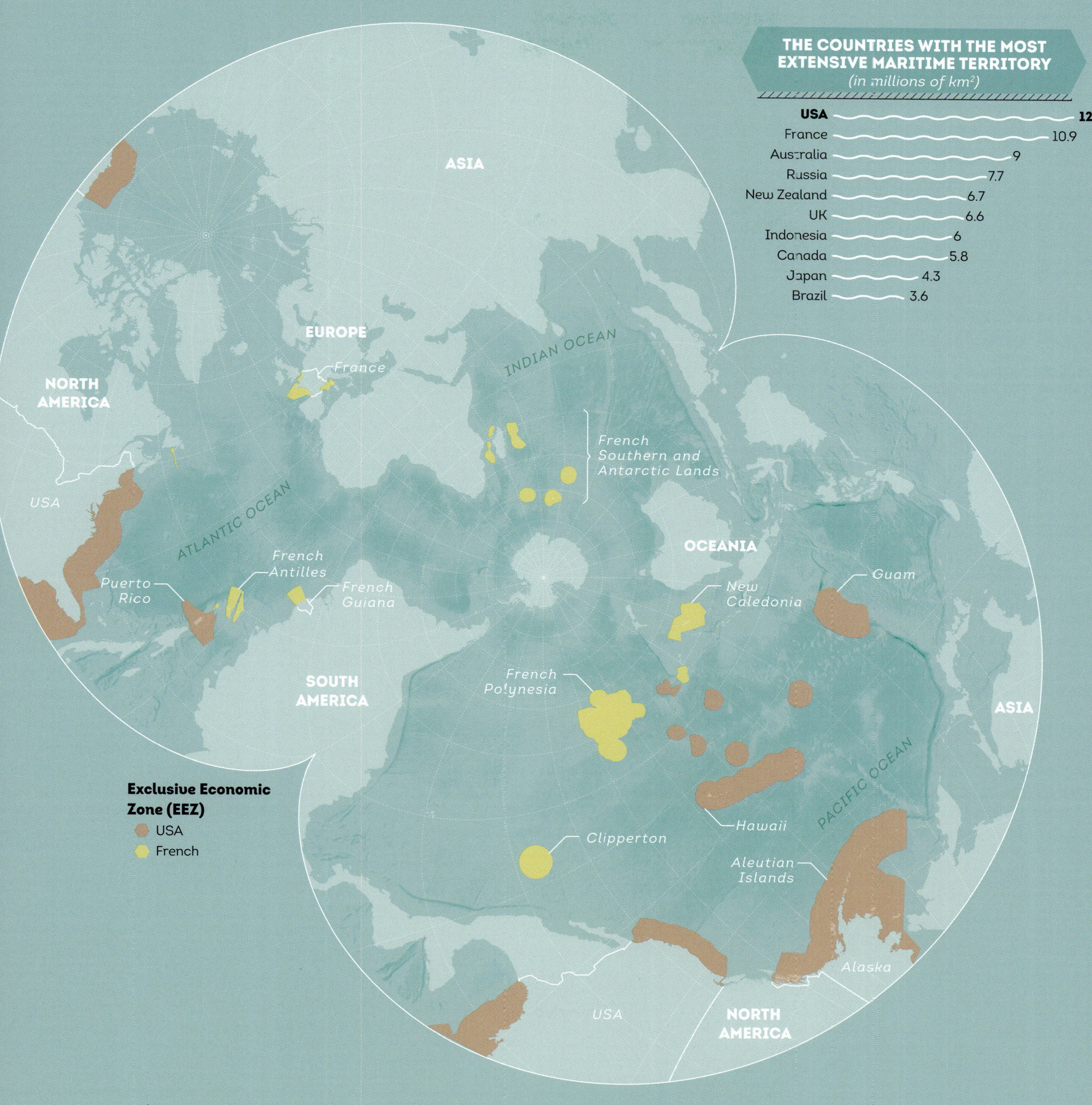

Sources: 'Géopolitique des îles en 40 cartes', *Le Monde*, special issue, 2019; Marineregions.org

12 THE ARCTIC

Claims from all sides

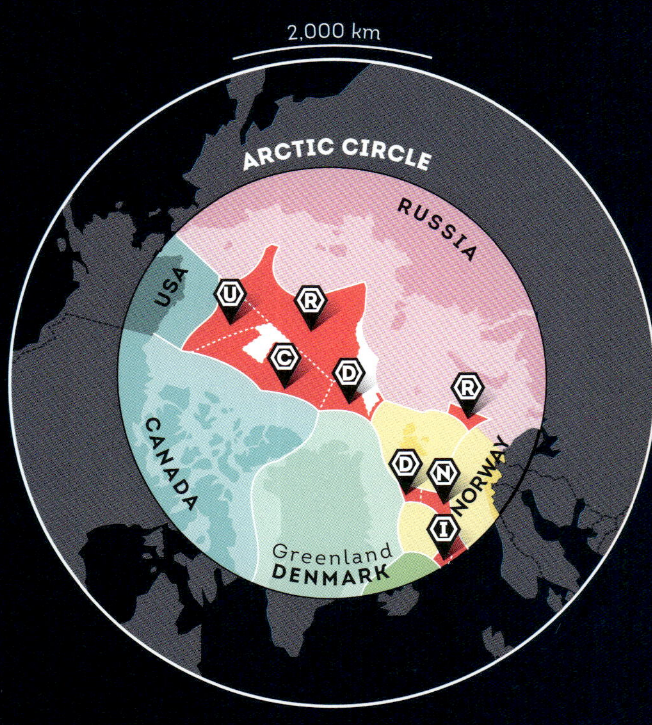

Territories and claims

Recognized border

Border of EEZ (200 nautical miles)

Exclusive economic zone (EEZ): Russ. Norw. Icel. Den. Can. USA

Continental shelf claims

USA

MILITARY PRESENCES AND CLAIMS

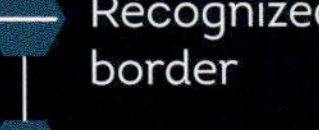
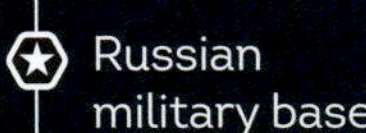

- Recognized border
- High seas (international zone)
- Area being claimed
- Russian military base
- US military base
- Territorial disputes
- Potential or proven oil deposits

ROUTES THROUGH THE ARCTIC

- Northwest Passage
- Northeast Passage
- Northern Sea Route (Russian territorial waters)
- Proposed 'Transpolar Sea Route'

Sources: M. Mered, *Les Mondes polaires*, Paris: PUF, 2019; M. Foucher, *L'Arctique, la nouvelle frontière*, Paris: CNRS Éditions, 2014; Arctic Portal; E. Canobbio, *Atlas des pôles*, Paris: Autrement, 2017

SVALBARD

Islands with a unique status

The Norwegian territory of Svalbard (Spitsbergen is the largest island) has a unique status: it is governed by a treaty that dates back to 1920, which states that its signatories can freely exploit its resources but may not use it for military purposes. Although the treaty defines the area to which it applies (the Svalbard archipelago), changes to maritime law have made it difficult to delimit its boundaries. Norway's interpretation of the treaty and of the United Nations Convention on the Law of the Sea (UNCLOS) conflicts with that of other European states. It has effectively defined a 'fisheries protection zone' around Svalbard, plus a 'fishing zone' around the island of Jan Mayen. This has led to disputes around the fishing of snow crabs; Norway classes them as a sedentary species and thus claims the exclusive right to exploit this resource. However, a dispute between Russia and Norway was resolved in 2010, with the two countries delimiting their borders in accordance with UNCLOS, rather than the treaty. These delimitations leave two areas of 'high seas' or international waters, known as the 'Banana Hole' and the 'Loophole', within national waters.

THE MEDIAN LINE

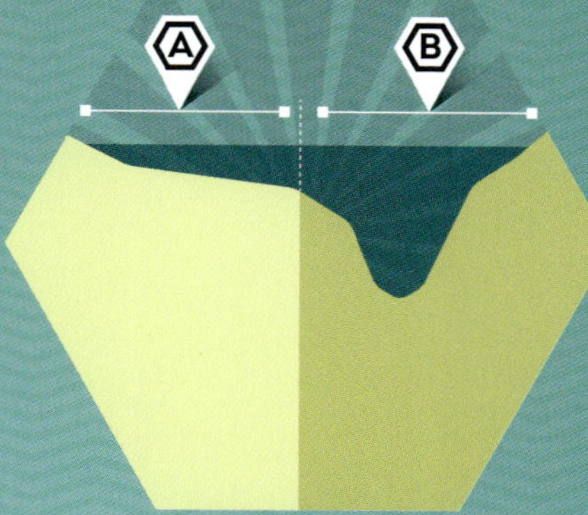

The border runs **equidistant from both banks**. The line is easy to draw but the water depth is not taken into account. Both countries have an equal share of the surface, but the volume of water belonging to each country may be different.

Example
The Bidassoa river between France and Spain.

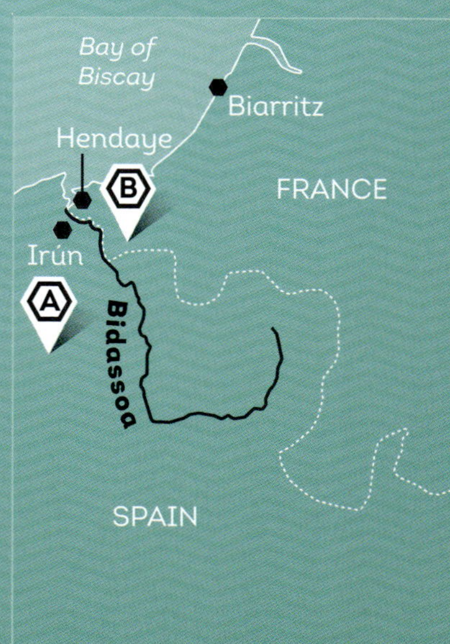

THE THALWEG

The border follows **the thalweg**, which is a **line joining the lowest points on the river bed**. This method has been favoured since the 19th century, as it makes shipping easier by taking the depth of the river into account.

Example
The Shatt al-Arab border between Iraq and Iran follows the thalweg, so both countries can use the river for shipping. This became one of the causes of the Iran-Iraq War.

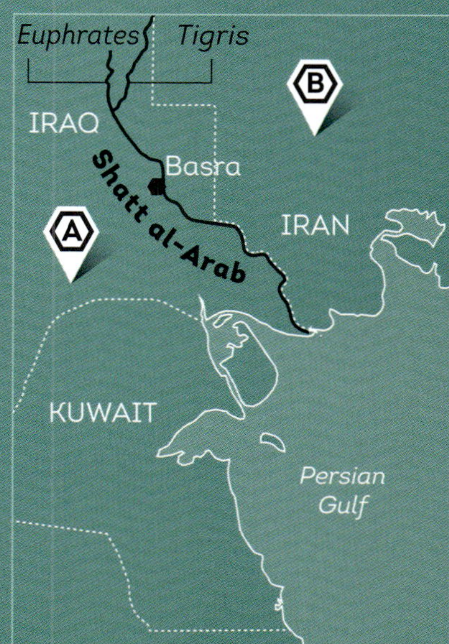

TRANSBOUNDARY RIVER

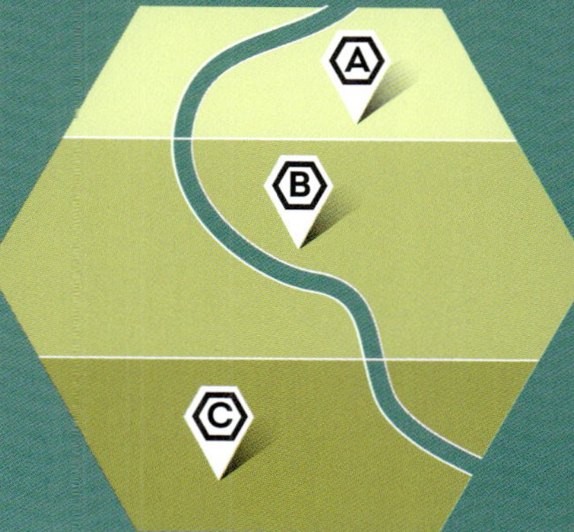

The waters of a river that **runs through multiple states** are shared. The countries downstream are dependent on those upstream.

Example
The Euphrates has its source in Türkiye and runs through Syria, then Iraq. Turkish hydroelectric stations have considerably reduced the downstream flow.

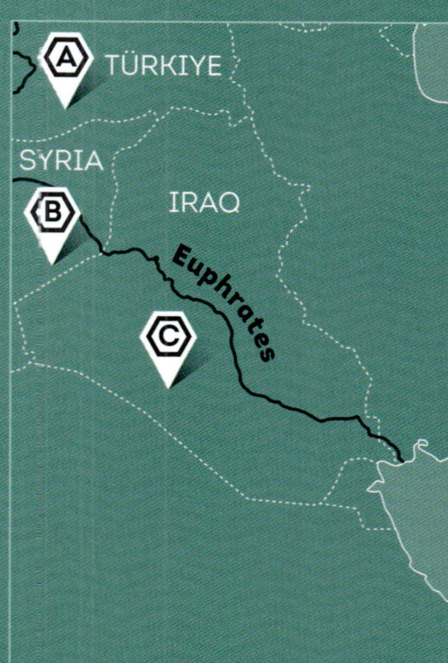

Sources: François Schroeter, 'Les Systèmes de délimitation dans les fleuves internationaux', *Annuaire français de droit international*, Paris: CNRS Éditions, no. 38, 1992/1; Mutoy Mubiala, 'L'Évolution du droit des cours d'eau internationaux', Graduate Institute Publications, 1995 [online].

17

THE EASTERN MEDITERRANEAN

Where natural gas is powering disputes

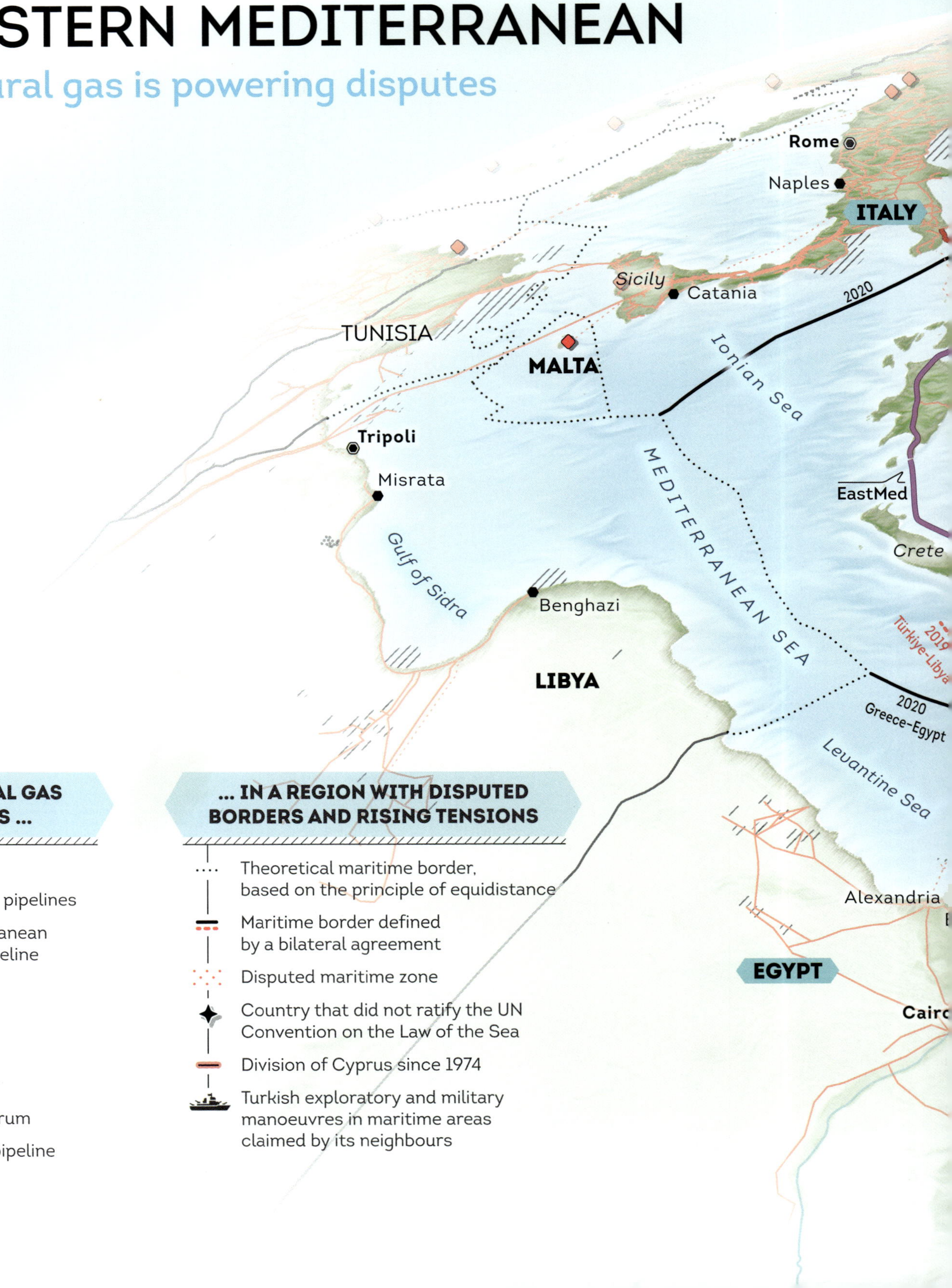

THE PRESENCE OF NATURAL GAS IS STOKING RIVALRIES ...

- Major gas reserves
- Gas pipelines - - - - Planned pipelines
- Proposed Eastern Mediterranean (EastMed) transnational pipeline
- Potential gas reserves
- Major terminal for import and export of liquefied natural gas (LNG)
- XXX Member states of the East Mediterranean Gas Forum
- Existing transnational gas pipeline

... IN A REGION WITH DISPUTED BORDERS AND RISING TENSIONS

- Theoretical maritime border, based on the principle of equidistance
- Maritime border defined by a bilateral agreement
- Disputed maritime zone
- Country that did not ratify the UN Convention on the Law of the Sea
- Division of Cyprus since 1974
- Turkish exploratory and military manoeuvres in maritime areas claimed by its neighbours

Sources: *Le Monde*, 27 September 2020; D. Ortolland & J.P. Pirat, *Geopolitical Atlas of the Oceans*, Patris: Technip, 2017 Marineregions.org; Gas Infrastructure Europe

ISRAEL AND LEBANON

An historic agreement on maritime borders

In October 2022, US mediation allowed Israel and Lebanon to resolve a dispute over their maritime border, in an area rich in natural gas, known as the 'Qana Prospect'. Israel claimed that the northern boundary of its exclusive economic zone ran along Line 1, while Lebanon claimed that the southern boundary of its EEZ ran along Line 23, a position eventually accepted by Israel, because stabilizing the situation would make it easier for Israel to exploit another gas field situated a little to the south (Karish). Although it is sometimes seen as a first step towards the normalization of relations between the two countries, the agreement – which was approved by Hezbollah – should not be seen as signalling a rapprochement, and leaves the issue of the land border unresolved.

HISTORIC AGREEMENT ON THE ISRAEL–LEBANON MARITIME BORDER

After the discovery of natural gas reserves in the Eastern Mediterranean in 2010 ...

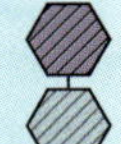

Gas reserves discovered in 2010

Other reserves

Agreement between coastal states on their maritime border

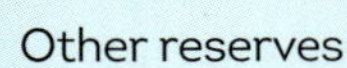

Fishing zone (2009), defining the area where Gazans have the right to fish

... although Lebanon and Israel have been in conflict for 75 years ...

Point 1B, marked in the 1949 armistice agreement and adopted by Lebanon as a demarcation of its border, but not recognized by Israel

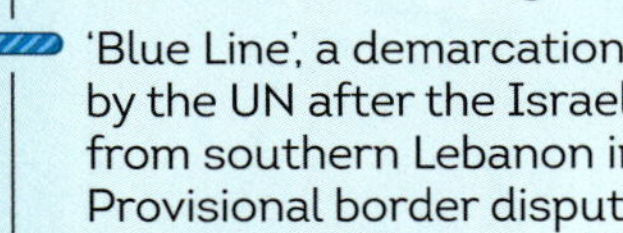

'Blue Line', a demarcation established by the UN after the Israeli retreat from southern Lebanon in 2000. Provisional border disputed by Lebanon and monitored by the UN

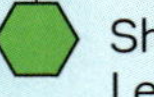

Shebaa Farms, an area claimed by Lebanon (called Mount Dov by Israel)

... Lebanon, under economic pressure, agreed to sign an agreement on its maritime border with Israel

Maritime border claimed by Lebanon in 2011 in 2020

Maritime border claimed by Israel

Proposition from US mediators in 2012, accepted by both parties on 11 October 2022

Rosetta Mouth

Alexandria

Nile Delta

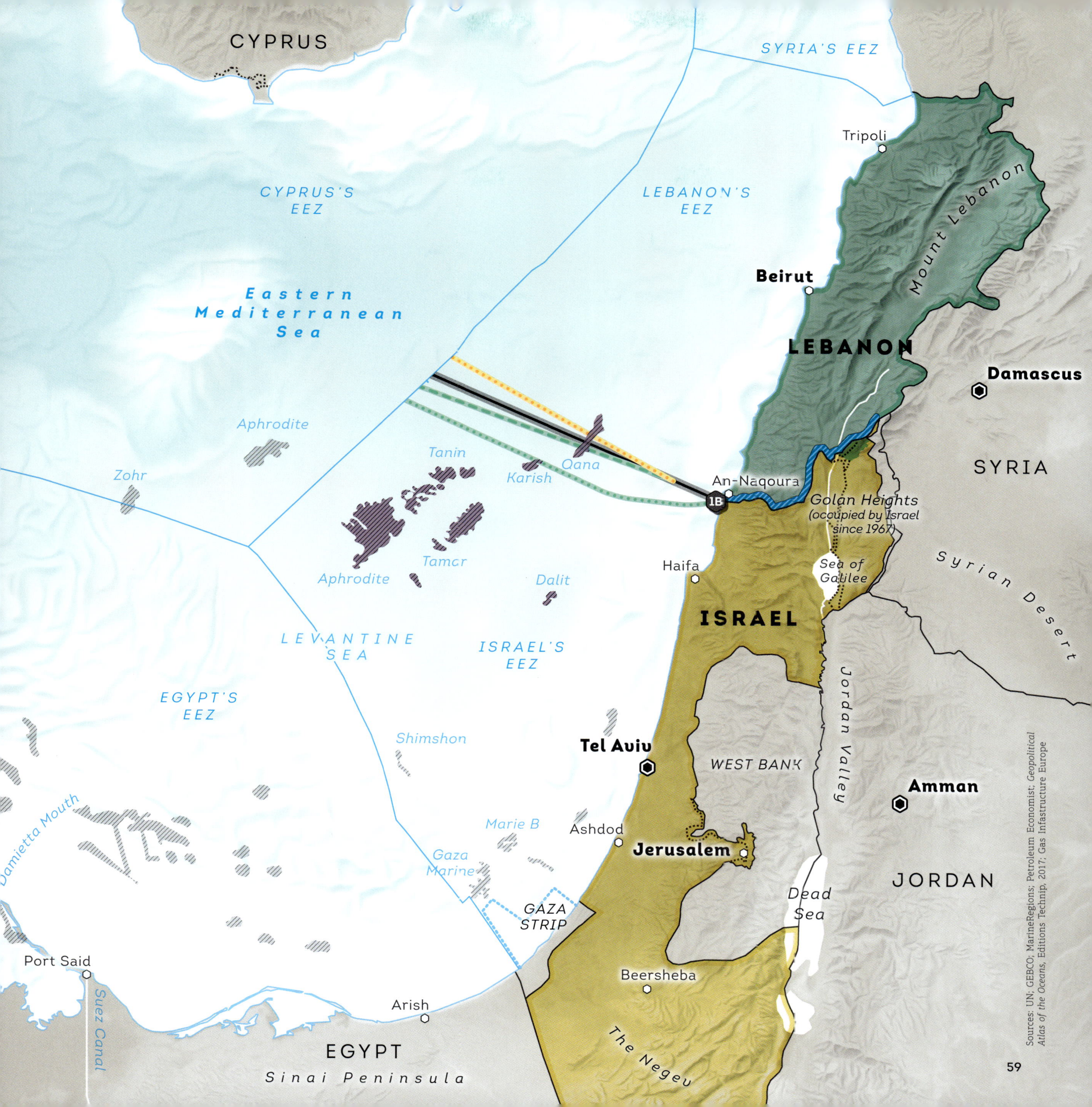

Sources: UN; GEBCO; MarineRegions; Petroleum Economist; *Geopolitical Atlas of the Oceans*, Editions Technip, 2017; Gas Infrastructure Europe

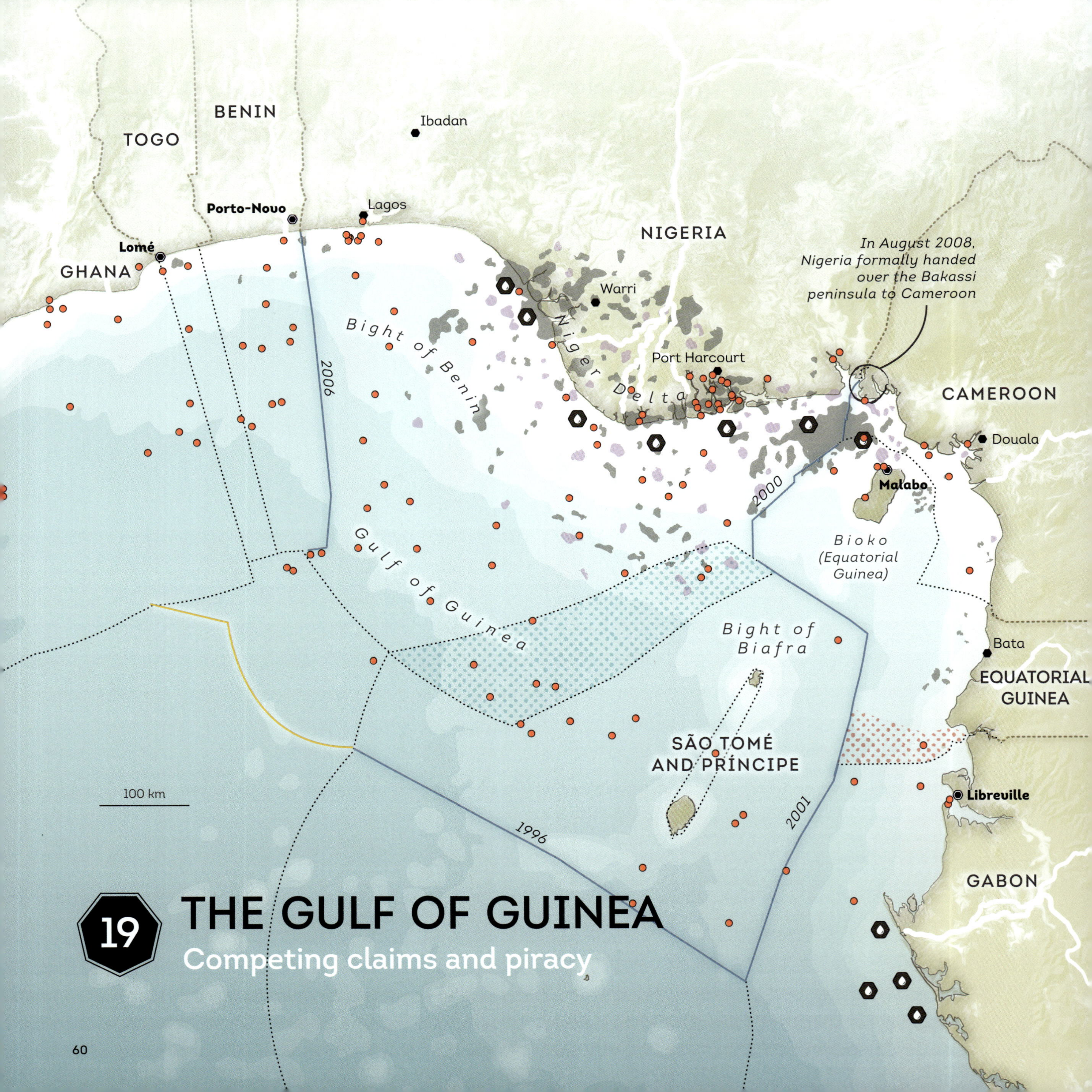

19 THE GULF OF GUINEA

Competing claims and piracy

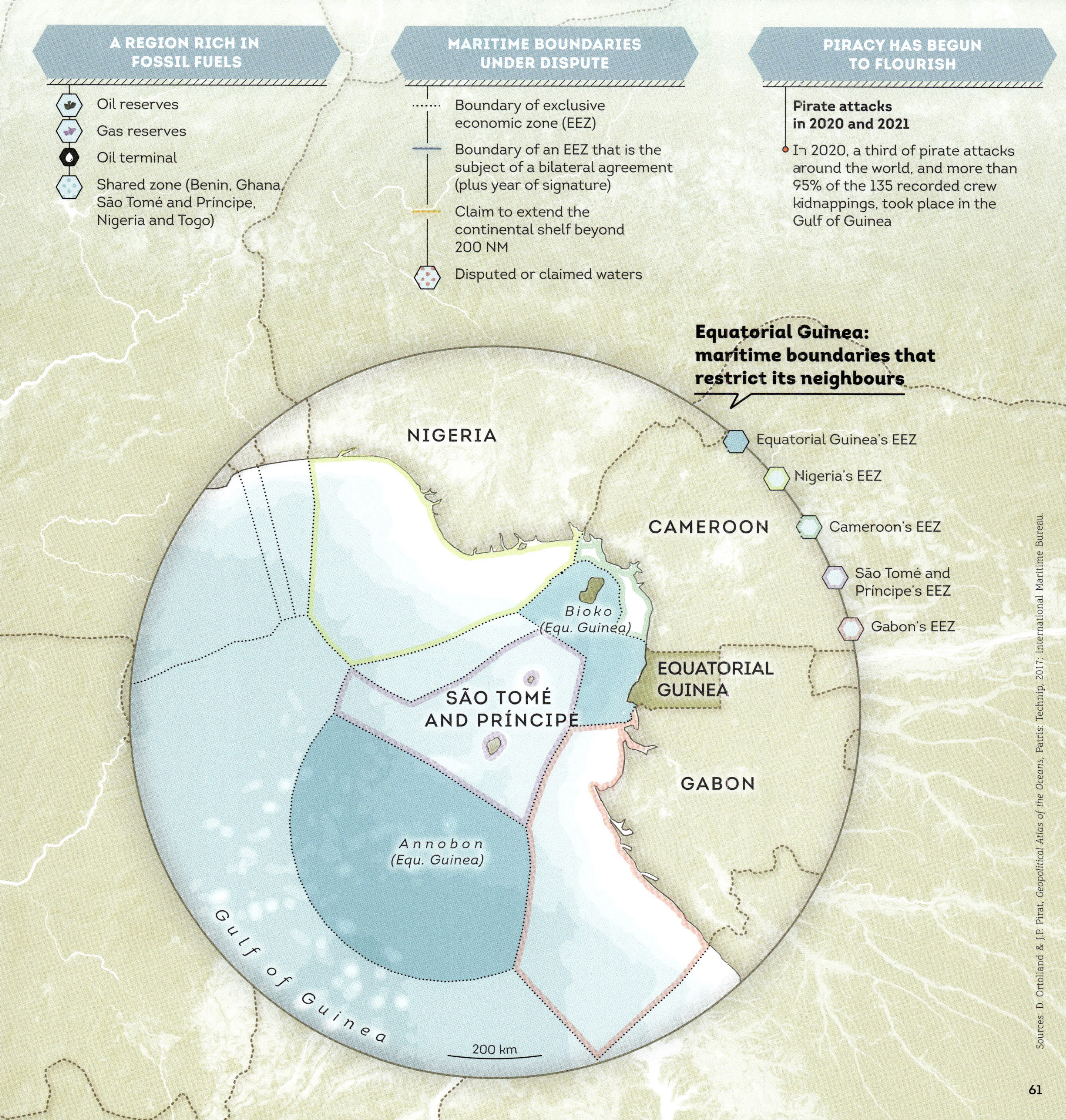

Sources: D. Ortolland & J.P. Pirat, *Geopolitical Atlas of the Oceans*, Patris: Technip, 2017; International Maritime Bureau.

RUSSIA
Sea of Okhotsk
Kamchatka Peninsula
South Kurils/ Northern Territorie
Vladivostok
Sea of Japan
Beijing
NORTH KOREA
Dokdo/Takeshima (Liancourt Rocks)
South Korea ADIZ
Seoul
SOUTH KOREA
Yellow Sea
JAPAN
Tokyo
Pacific Ocean
Shikoku
East China Sea
Shanghai
CHINA
China ADIZ
Japan ADIZ
Bonin Islands
Senkaku/ Diaoyu
Ryukyu Islands
Okinawa
Taipei
TAIWAN
Sakashima Islands
Volcano Islands
South China Sea
Taiwan ADIZ
Okinotorishima
500 km
PHILIPPINES

THE OUTERMOST ISLANDS OF JAPAN

An archipelago with disputed borders

The Kuril Islands have been claimed by both Russia and Japan since the 19th century. The Treaty of St Petersburg (1875) granted the islands to Japan. They came under Soviet control in 1945. Japan still claims the four islands that lie nearest to it – known as the Northern Territories – and does not consider them part of the Kuril Islands. In 2022, Russia withdrew from bilateral negotiations. Meanwhile, Japan disputes South Korea's administration of the Liancourt Rocks (known as Dokdo in Korean, and Takeshima in Japanese). Lying halfway between their territories, these islands are at the centre of a dispute that stirs up nationalist sentiment in both countries. Japan's dispute with China is centred on the Senkaku/Diaoyu Islands (also claimed by Taipei), whose waters are rich in natural resources. These islands are under US protection. Japan wants to apply the principle of equidistance, but China claims that its continental shelf reaches as far as the islands and has made a number of intrusions into these disputed waters.

Minamitorishima

SEA BORDERS

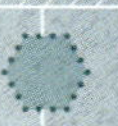
Japanese exclusive economic zone (EEZ)

Maritime zone claimed by Japan

Disputed maritime areas and territories

Under Russian administration, claimed by Japan

Under South Korean administration, claimed by Japan

Under Japanese administration, claimed by China and Taiwan

US ally

US bases and facilities

Countries with common interests

AIR BORDERS

Air Defence Identification Zone (ADIZ)

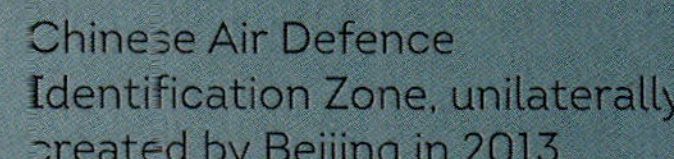
Chinese Air Defence Identification Zone, unilaterally created by Beijing in 2013

21
THE SOUTH CHINA SEA
A complex game of strategy
Taipei
TAIWAN
Hong Kong
Pratas Island
CHINA
MYANMAR (BURMA)
LAOS
HAINAN
Paracel Islands
Macclesfield Bank
Scarborough Shoal
LUZON
PHILIPPINE SEA
THAILAND
SOUTH CHINA SEA
Manila
CAMBODIA
Spratly Islands
VIETNAM
PHILIPPINES
GULF OF THAILAND
PALAWAN
SULU SEA
MINDANAO
ANDAMAN SEA
James Shoal
MALAYSIA
BRUNEI
MALAYSIA
CELEBES SEA
Singapore
STRAIT OF MALACCA
BORNEO
SUMATRA
INDONESIA
SULAWESI
INDONESIA
500 km

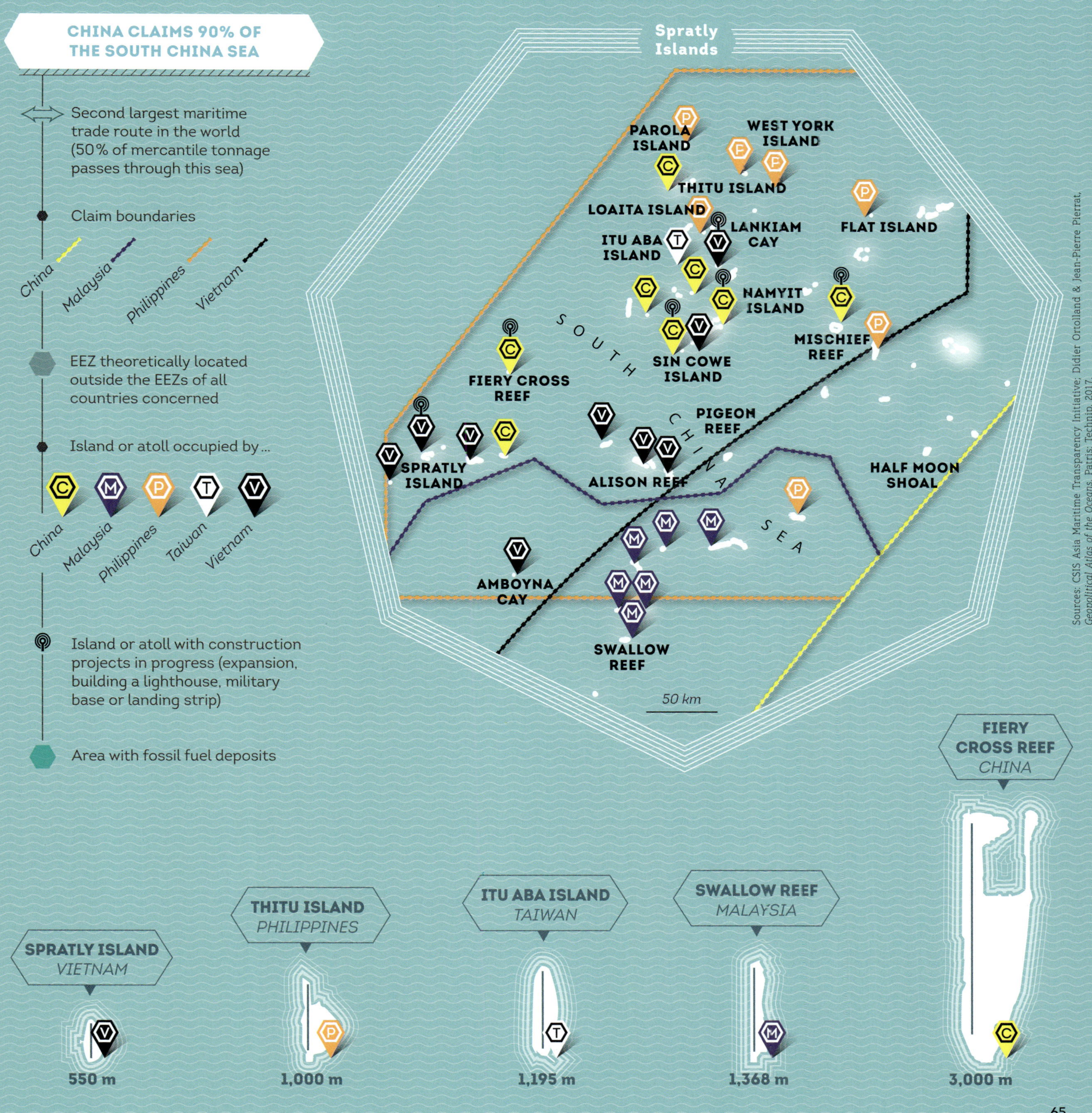
CHINA CLAIMS 90% OF THE SOUTH CHINA SEA
Second largest maritime trade route in the world (50% of mercantile tonnage passes through this sea)
Claim boundaries
China
Malaysia
Philippines
Vietnam
EEZ theoretically located outside the EEZs of all countries concerned
Island or atoll occupied by ...
China
Malaysia
Philippines
Taiwan
Vietnam
Island or atoll with construction projects in progress (expansion, building a lighthouse, military base or landing strip)
Area with fossil fuel deposits
Spratly Islands
PAROLA ISLAND
WEST YORK ISLAND
THITU ISLAND
LOAITA ISLAND
LANKIAM CAY
FLAT ISLAND
ITU ABA ISLAND
NAMYIT ISLAND
MISCHIEF REEF
SIN COWE ISLAND
FIERY CROSS REEF
SOUTH CHINA SEA
PIGEON REEF
SPRATLY ISLAND
ALISON REEF
HALF MOON SHOAL
AMBOYNA CAY
SWALLOW REEF
50 km
FIERY CROSS REEF
CHINA
SPRATLY ISLAND
VIETNAM
THITU ISLAND
PHILIPPINES
ITU ABA ISLAND
TAIWAN
SWALLOW REEF
MALAYSIA
550 m
1,000 m
1,195 m
1,368 m
3,000 m
Sources: CSIS Asia Maritime Transparency Initiative; Didier Ortolland & Jean-Pierre Pierrat, Geopolitical Atlas of the Oceans, Paris: Technip, 2017.

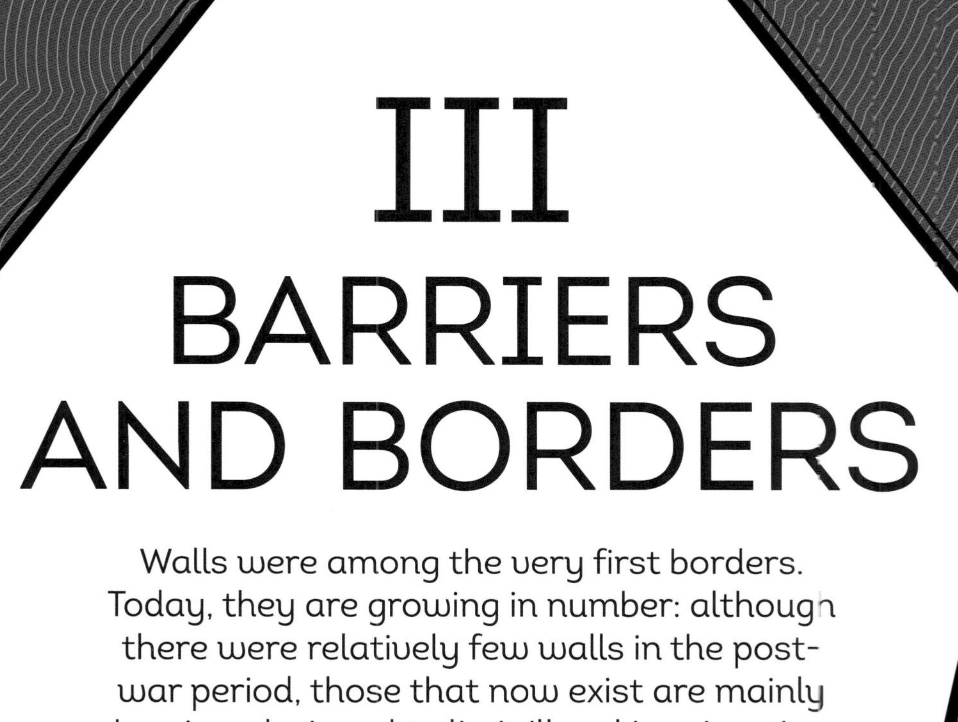

III
BARRIERS AND BORDERS

Walls were among the very first borders. Today, they are growing in number: although there were relatively few walls in the post-war period, those that now exist are mainly barriers designed to limit illegal immigration and smuggling, as well as infiltration by terrorist groups, which remains a rising source of fear.

III · BARRIERS AND BORDERS

It is rare for borders to be marked by a wall in the strictest sense of the word. But the term is widely used to refer to a physical barrier that can be anything from a simple wire fence to a proper wall. Today, only a small proportion of borders have barriers of this kind: depending on your definition of a wall and how you calculate it, the figure is between 6% and 18% of land borders. Around 70 walls or barriers – mainly in the Middle East and Europe – have been built at around 80 dyad borders, representing a quarter of all bilateral land borders. In some cases, there are 'virtual walls': highly advanced surveillance (sensors, drones, satellites) but no physical barrier (for example, Brazil). These modern forms of border control come under the term 'smart borders'.

Walls are highly symbolic, and there are a number of well-known historical examples: the Great Wall of China, Hadrian's Wall, and of course the Berlin Wall. At first the latter was nothing more than a barbed-wire fence erected in a single night, but it went on to completely surround West Berlin, stretching a total of 165 km (with 106 km of wall) for 28 years, monitored by 14,000 guards and 6,000 dogs. The first known border wall was built in around 2040 BCE, between the Tigris and the Euphrates, by the Sumerian king Shulgi.

The history of the Great Wall of China stretches back two thousand years. Strictly speaking, there were a number of 'great walls': a series of imperial walls built in the north of the country, stretching a total of 50,000 km. The first Walls of China were built by the Emperor Qin Shi Huang in the 3rd century BCE; firstly to stop his army from being pushed back, then to control trade. The Chinese walls then became barriers to protect against invasion, not designed to be impenetrable but instead to hold the enemy, the 'barbarians', in one place. It wasn't until the 15th century that the Ming Great Wall was built (3,640 km, with more than 2,860 km of fortifications) – this is the wall that still partly remains standing today.

Hadrian's Wall and the Antonine Wall (both 2nd century CE) in the UK are equally well known, but there are many other examples of major imperial works: the Gates of Alexander, which stretched from the Caspian Sea to the Caucasus (6th century), Sungbo's Eredo (9th century) in Nigeria, and 'Trajan's Wall' in medieval Eastern Europe.

The Berlin Wall was very unusual: it was one of the few known walls – along with the Iron Curtain overall, and the North Korean border today – designed to stop people from leaving, rather than entering.

A proliferation of walls, barriers and fences

There were only around fifteen border walls and fences at the end of the Cold War; today there are more than 70, including twenty or so solid walls. They are being built everywhere: in southern Africa, Central Asia, South Asia. The US, Israel, Saudi Arabia and India, among others, have plans to build barriers along most of their borders – the longest fence in the world separates India from Bangladesh (3,200 km). Since 9/11, terrorism has been a major reason for this acceleration. But it should also be seen in the context of globalization and the increase in movement, both legal and illegal, of people and goods. The purpose of a wall may change over time. The barrier built by Botswana on its border with Zimbabwe was originally designed to control the movement of wild animals; it has now become a defence against immigration and trafficking. The Sahara Wall was initially intended to protect Moroccan interests against the Sahrawi rebellion: now, its ability to control migration is equally important. The same is true of the US–Mexico wall: it was originally designed to stop migrants, but it also plays a key role in the fight against drug trafficking. The reverse is true in Israel: the barrier that separates it from Sinai was originally designed as an anti-terrorist measure, and now it is used to prevent immigration from Africa. Walls also create de facto borders (not legal ones): a practice known as 'borderization', as seen in Kashmir, Israel, Georgia, etc.

Walls may be underground (Gaza) and one day they may even float in the sea (a Greek project). They typically include buffer zones: 9 km between Iraq and Kuwait (6 km of which are in Iraqi territory), up to 7 km in Cyprus, 5 km in the Western Sahara (on the Moroccan side), 4 km on the Korean border (two on each side, with a Joint Security Area that stretches for 800 metres). There is a similar situation around Gaza, with the Philadelphi Corridor on the Egyptian side (200 to 300 metres) and a strip of a few dozen metres on the Israeli side. One exception is the border between India and Bangladesh: the 150-metre buffer zone is inhabited, making life tough for its residents.

While walls are popular with politicians – building one is seen as a simple, visible and 'common sense' response to border issues – they are far less so among philosophers. Thierry Paquot believes 'a wall expresses a lack of understanding, separation, segregation [...] The builder of walls is a polluter of humanity.' For fellow philosopher Wendy Brown, walls 'contribute new forms of xenophobia and parochialism to a postnational era.' For legal expert Monique Chemillier-Gendreau, 'they separate humans from one another and are a violation of the *jus communicationis* between humans.'

Are walls ineffective, as is often claimed? Former Secretary of Homeland Security Janet Napolitano said: 'Show me a 50-foot wall and I'll show you a 51-foot ladder.' There is no easy answer. Of course, closed borders are regularly crossed. Digging tunnels is commonplace, although this has only been systematically done in a few cases (Mexico, Gaza, North Korea). The Egyptian wall around Gaza was even blown up by Hamas in 2008. But walls do slow down and channel the flow of people. In Israel, they have made it possible to considerably reduce attacks. In the US, illegal immigration has decreased, with border patrols arresting around half of those who attempt to cross.

Walls are popular with politicians – building one is seen as a simple, visible and 'common sense' response to border issues.

But closing borders can have many negative effects: it forces immigrants to take physical risks; it turns seasonal migration into permanent emigration; it decreases economic growth; it exacerbates bilateral disputes; and it also blocks the natural migration patterns of land animals.

Walls are also places where artistic expression flourishes, as was formerly the case in Berlin, and now in Mexico and the West Bank. They are the setting for political theatre and ceremonies, e.g. in the Joint Security Area between North and South Korean, or at the border crossings in Wagah in Pakistan, or Petrapole in India. They can even become places where nature thrives, like the Korean DMZ.

Anti-migrant walls

We live in a time of unprecedented movement across borders. The growth of road, rail and air transport has given humans an exceptional level of mobility. The term 'migration' refers to long-term expatriation, which may be temporary or permanent (cross-border working doesn't fall under this category). Migration can be chosen – people moving abroad for their studies or to have new experiences – or forced – due to war, oppression, unemployment or poverty. In 2021, there were 281 million migrants around the world, 36 million of whom were refugees.

Up until the 19th century, it was easier to enter national territory than to leave it. People often wanted to leave in order to escape conscription or taxes. The right to leave national territory was enshrined in the Universal Declaration of Human Rights (1948). But today, the situation is reversed.

When emigration is motivated by economic factors, it is the result of 'supply' (which depends on local circumstances and an individual's ability to move elsewhere) and 'demand' (perceived employment opportunities, whether legal or not, and the chance to join a community or family members). In strictly economic terms, immigrants are more often a benefit than a drain, as they rarely retire in their adopted country. They also contribute to growth in their country of origin: sending home funds totalling $500 billion a year, more than double governmental development aid. 'Migradollars' are the second most important source of income in Mexico after oil, totalling $20 billion a year.

The largest migrant flows are firstly intracontinental (Europe, Africa, Asia). When they are intercontinental, the direction of travel tends to be towards North America and Europe.

Borders first began to be systematically monitored around the time of the First World War. Passports became standardized. After the Second World War, visas appeared. While it became easier to physically move from one country to another, from a legal perspective, it became much more difficult. And there is a level of inequality when it comes to ease of movement.

Over the last twenty years or so, there has been increased externalization of borders by developed countries, which has been dubbed a 'return to the Limes' (Michel Foucher). Visas are checked by airlines in the country of departure. The country of origin or transit is required to cooperate on controlling illegal immigration (shared patrols, detention centres, repatriation agreements). The US checks goods being sent to America in foreign ports. Because of a bilateral agreement, the UK has border checks in Calais. In the European Union (EU), there are 60 million immigrants (born in a different country from their

We live in a time of unprecedented movement across borders. The growth of road, rail and air transport has given humans an exceptional level of mobility.

country of residence), which represents 11.7% of the population, two thirds of whom were born in a non-member state. This category partially overlaps with that of foreigners (whose nationality is not that of their country of residence); they number 40 million, 7.8% of the EU's population. More than half of these foreigners are non-European (including 9 million Africans, the largest number being Moroccans). It is 'small' countries (Luxembourg, Cyprus, Lithuania, Estonia, Austria, Ireland) that contain the largest proportion of foreigners and people born abroad. But in absolute figures, it is naturally the 'large' countries that have the most foreigners: Germany, the UK, France, etc.

Between 2015 and 2017, there were 3 million first-time asylum claims, mostly due to conflict (Afghanistan, Iraq, Syria). Today, the number of illegal immigrants in Europe, not including those with asylum claims being processed, is estimated at around 3 million.

The Schengen Agreement (1985) has 27 members: 23 EU member states and 4 non-member states. It was primarily designed to make cross-border trade and working easier; 1.7 million Europeans go to a neighbouring country every day to work. It includes a number of clauses:

- removing checks at internal borders. These can be reinstated in exceptional circumstances (safeguard clause, used a number of times since 1995);
- checks can be carried out in an area stretching 20 km on either side of the borders (as well as at airports and ports of entry), and there is stronger cooperation (Maastricht Treaty, 1992) around customs, law and policing;
- stronger checks at shared external borders, and shared monitoring systems (Treaty of Amsterdam, 1997): visas, asylum applications (Dublin II Regulation, 2003), databases (Eurosur system, 2013).

A dedicated agency, Frontex, was created to manage Europe's borders in 2004. Since 2007, its Rapid Border Intervention Team (RABIT) has been able to rapidly deploy border guards at the request of a member state in times of crisis. Frontex has become the European Border and Coast Guard Agency, and employs around 1,000 people. The Schengen Area has 1,700 entry points and 7,289 km of land borders. Frontex runs a number of maritime, terrestrial and aerial surveillance operations, in cooperation with border states: Themis (Italy), Poseidon (Greece), Minerva and Indalo (Spain), and the Western Balkans.

At the same time, Europe has built more and more barriers to protect itself from illegal immigration and trafficking, and also, increasingly, from Russia. More than half of the countries in the EU have built walls: on the border with Morocco (Spanish enclaves) in the 1990s; on borders with Türkiye (Greece, Bulgaria) in the 2000s; and more recently on borders in central and northern Europe. In 2015, it was already possible to say that Europe would soon have 'more physical barriers on its national borders than it did during the Cold War' (*The Economist*). While they measured around 300 km ten years ago, these barriers now stretch for more than 2,000 km. The new walls in Europe are the consequence of the Schengen Agreement, the war in Syria and Russian aggression. On the other hand, the states that signed the Schengen Agreement have also dismantled some of their barriers, as Slovenia and Croatia have done. Anecdotally, it is notable that the readmission agreement signed by Germany and Switzerland in 1993 explicitly states that the birth of a child 'is equivalent to entering through an external border'.

The US started to reinforce its southern border – the most-crossed border in the world – when the North American Free Trade Agreement came into force, which brought an end to the policy of regularization of illegal immigrants. India has also been building barricades to keep out migrants from Bangladesh: it has already built more than 3,000 km of barriers along a border measuring 4,100 km. Zimbabwe's neighbours are also seeking to halt the exodus of its people.

Walls slow people down and channel them into specific areas, but they can't completely stop illegal emigration; people will always find other ways. The walling off of the Spanish enclaves in Morocco forced migrants to travel by sea or through the Balkans. Stricter monitoring by Greece has reduced immigration at its border, but led to an increase in the number of people arriving via Türkiye and Eastern Europe. Increased surveillance at major crossing points on the US–Mexico border has forced migrants to try and cross the southern deserts. But this can have dramatic consequences: the number of deaths through drowning (Mediterranean) and dehydration (Arizona) has risen significantly in recent years.

Walls after war

Walls built after a conflict may symbolize a ceasefire, giving physical form to a stalemate, or they may be created unilaterally, following the line of a de facto border. Therefore, they rarely run along an internationally recognized border between two states. That is the case for Korea (the oldest existing wall), the Israel–Lebanon border (in part), and the barrier wall between Kuwait and Iraq. On the other hand, the walls built by Morocco (Western Sahara), Türkiye (Cyprus), Egypt (Gaza), Israel (Palestinian Territories) and India (Kashmir) were intended both to create a border and to give it physical form. Behind walls, people settle and colonize. The UN is always present in some way or other. But when they are the result of a stalemate, borders are likely to be sources of conflict: between Israel and Lebanon, and India and Pakistan, serious incidents have led to the use of force on a large scale. In Korea, such incidents have not, so far, escalated into war.

Post-conflict walls are not necessarily synonymous with completely closed borders. It is possible – although difficult – to cross the walls in Israel and Cyprus. By contrast, the Algeria–Morocco border has been closed since 1994, although today the physical barrier only runs along certain stretches.

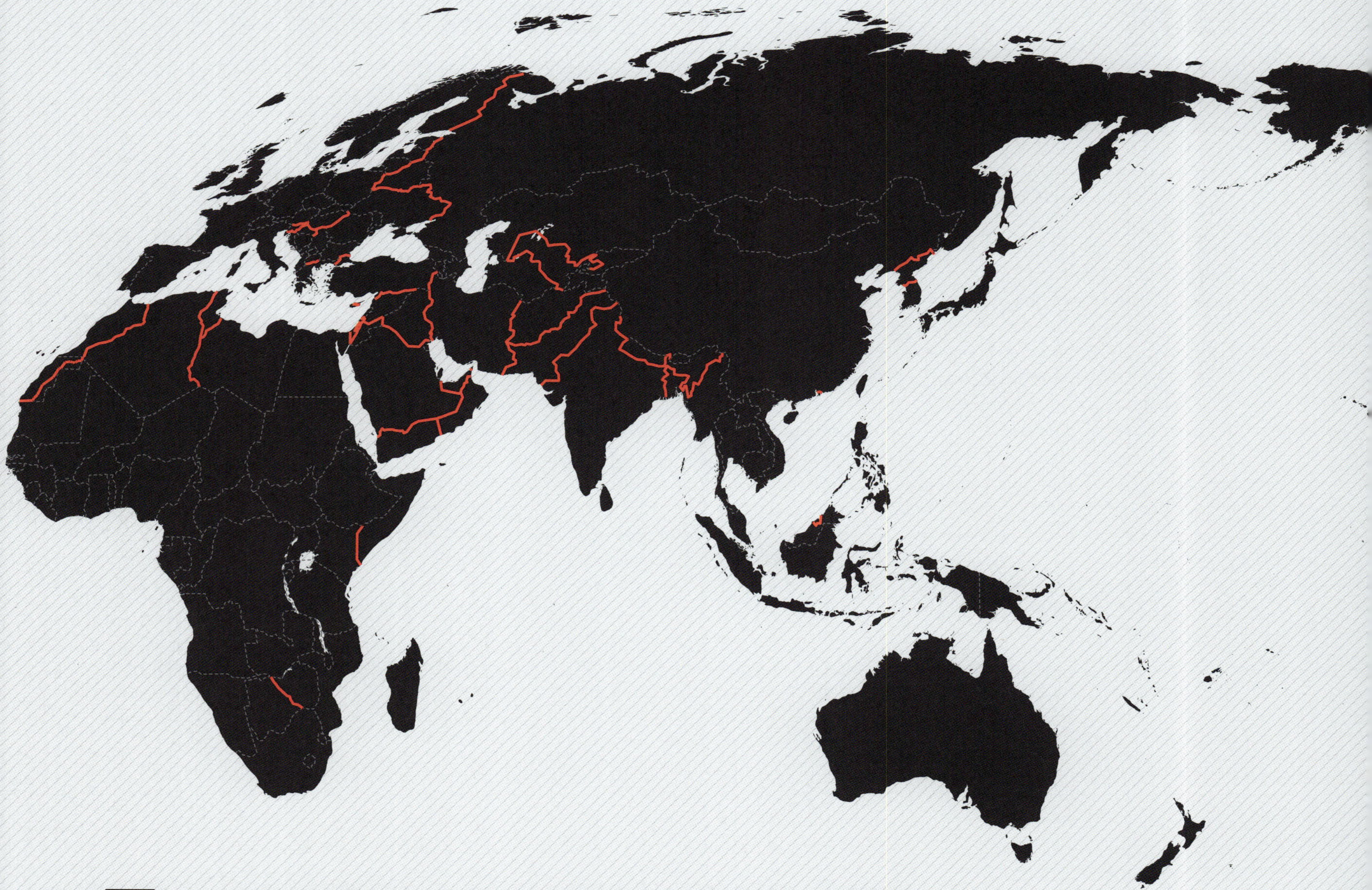

22 WHERE ARE THE WALLS?

Border barriers today

CURRENT BORDER WALLS AND FENCES, BY DATE OF CONSTRUCTION

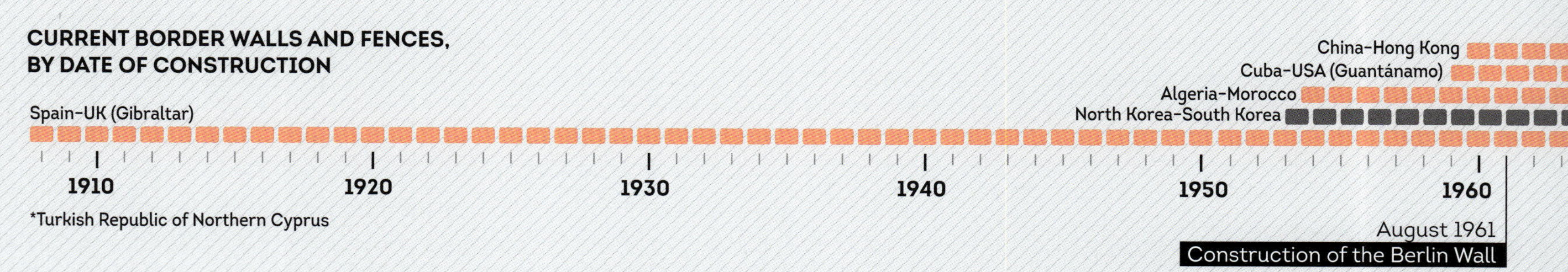

*Turkish Republic of Northern Cyprus

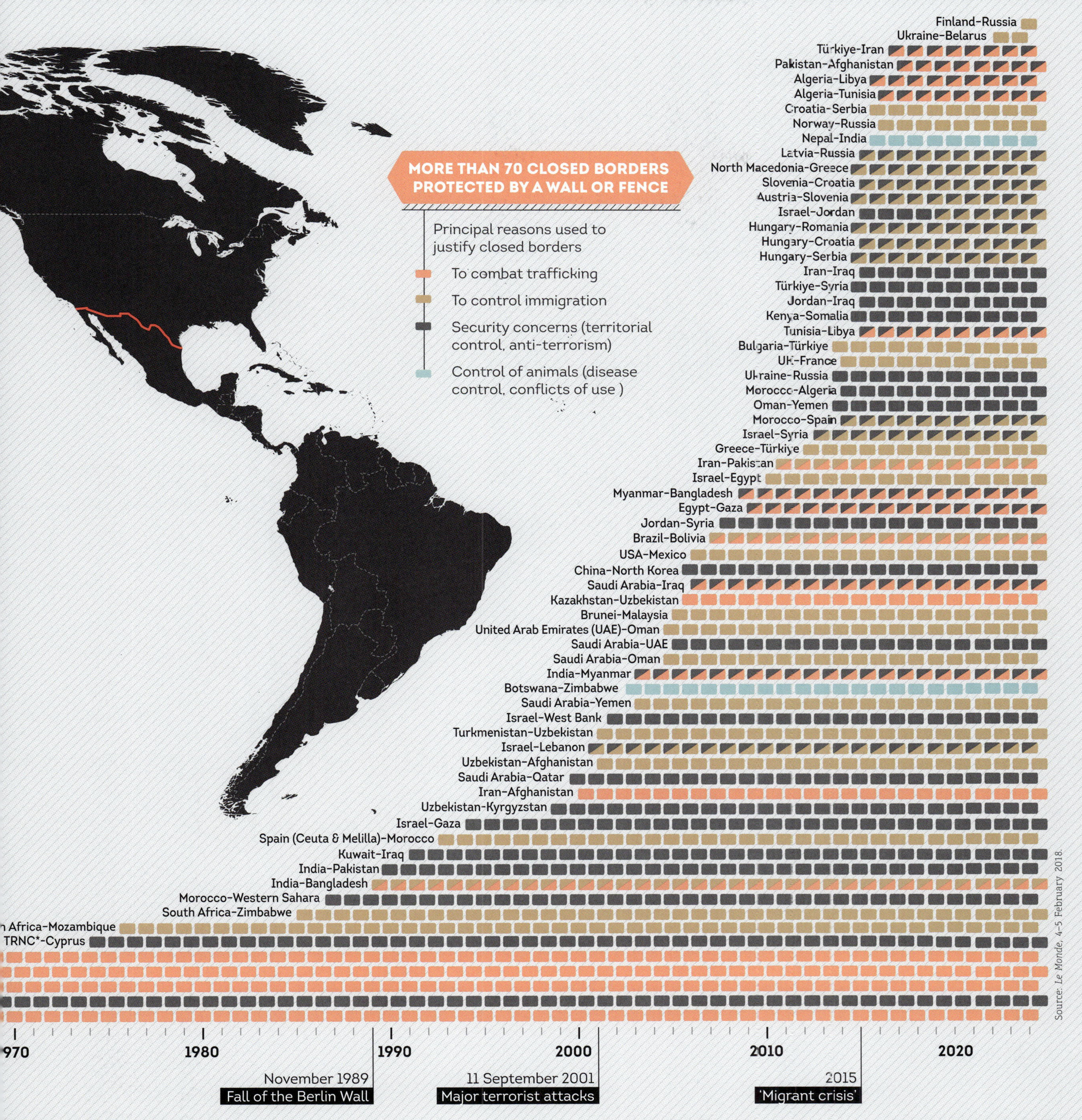

Source: *Le Monde*, 4–5 February 2018.

23 BUILDING A BARRIER

Four types of border wall

USA-MEXICO

A wall across North America

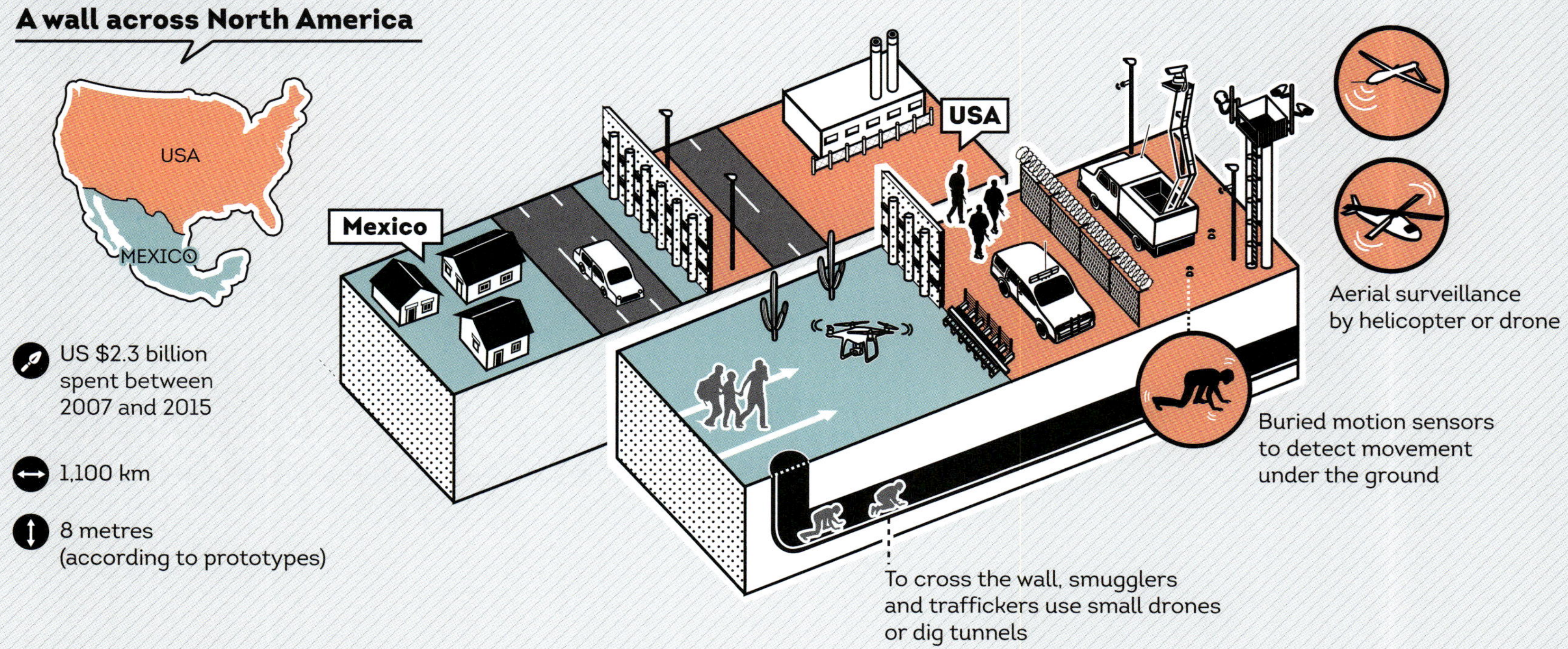

SPAIN-MOROCCO

Five fences form a boundary between Europe and Africa

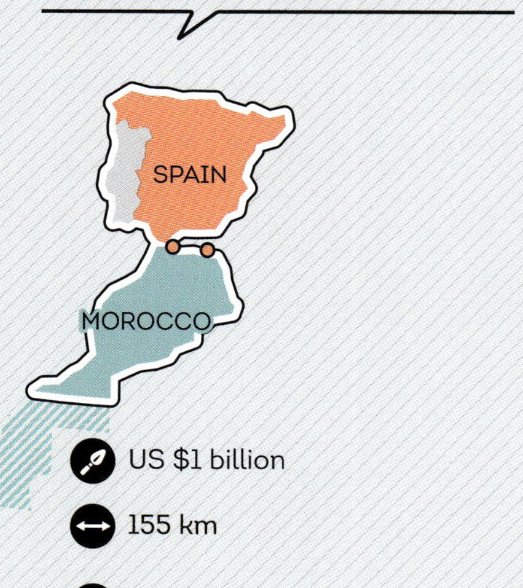

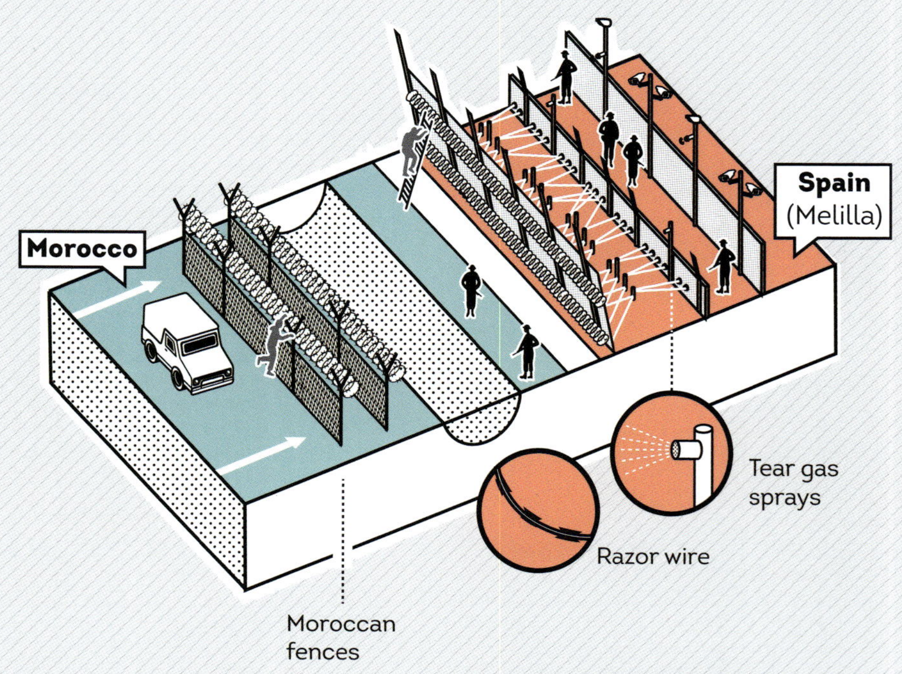

NORTH KOREA–SOUTH KOREA
The world's most militarized border

NORTH KOREA
SOUTH KOREA

More than US $100 million

248 km

DMZ
Military Demarcation Line (border)
North Korea
South Korea
2 km
2 km
Military zone (civilian access restricted)
Military zone (civilian access restricted)
Incursion tunnel built by North Korea (four tunnels discovered between 1974 and 1990)

HUNGARY-SERBIA
Eastern Europe tries to prevent migration

US $1 billion

155 km

4 metres

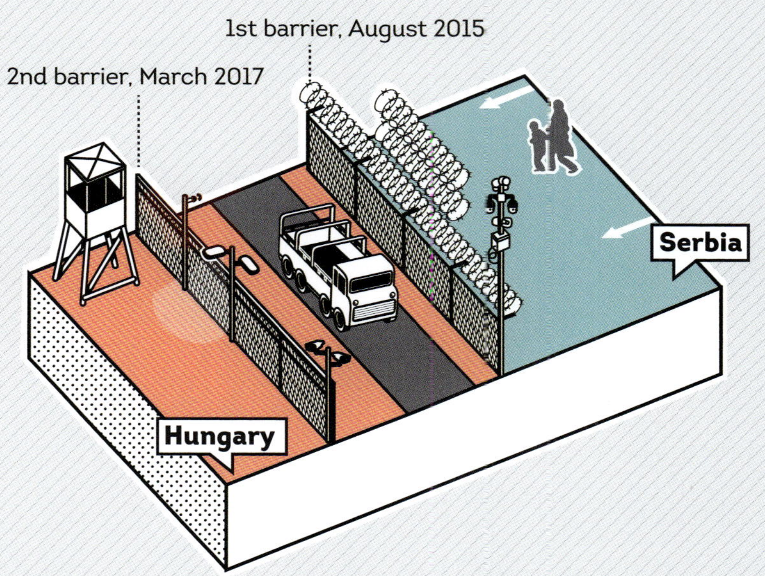

Source: *Le Monde*, 4–5 February 2018.

24 THE DEADLY COST OF MIGRATION

Lives at risk in the Mediterranean

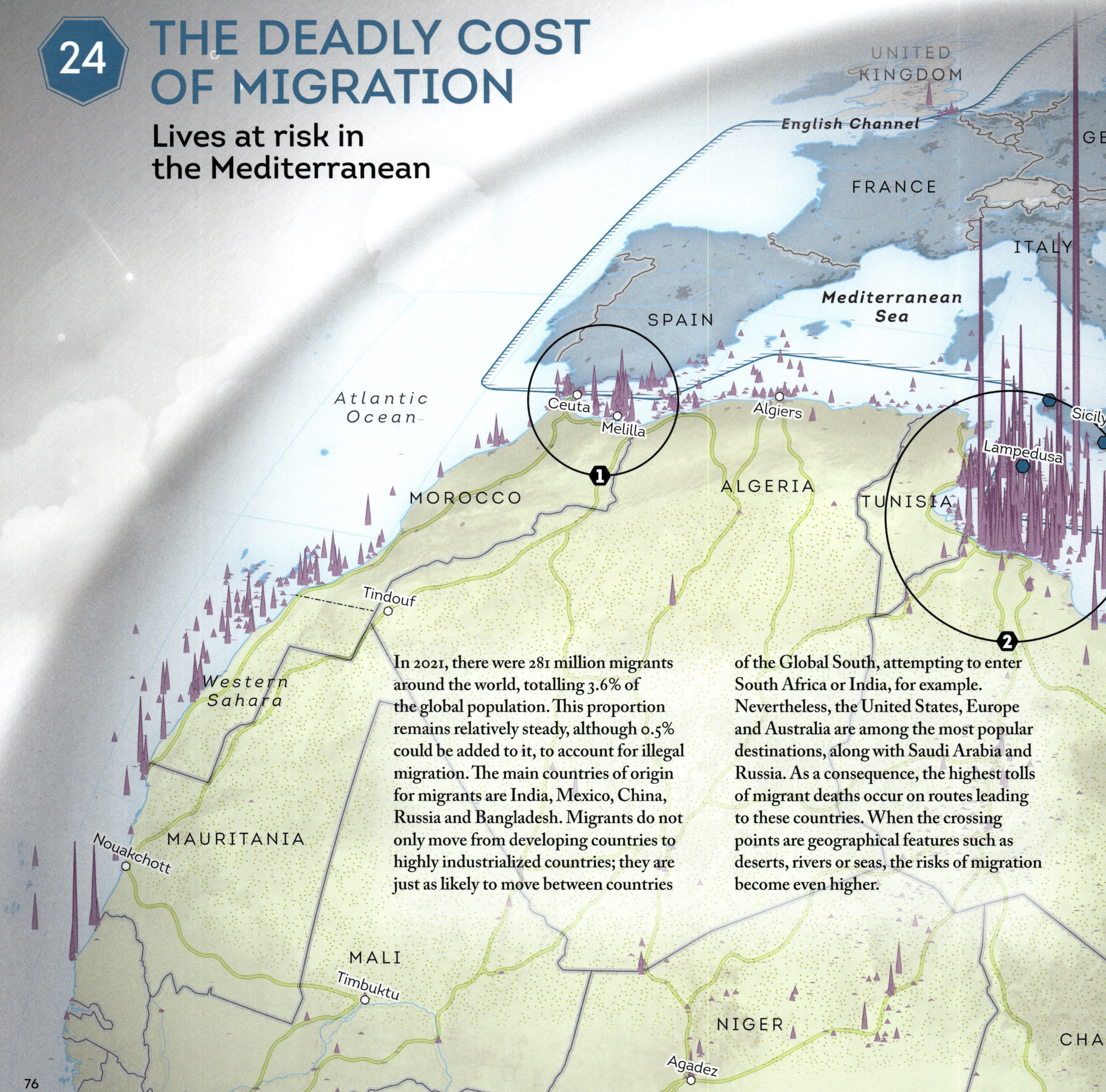

In 2021, there were 281 million migrants around the world, totalling 3.6% of the global population. This proportion remains relatively steady, although 0.5% could be added to it, to account for illegal migration. The main countries of origin for migrants are India, Mexico, China, Russia and Bangladesh. Migrants do not only move from developing countries to highly industrialized countries; they are just as likely to move between countries of the Global South, attempting to enter South Africa or India, for example. Nevertheless, the United States, Europe and Australia are among the most popular destinations, along with Saudi Arabia and Russia. As a consequence, the highest tolls of migrant deaths occur on routes leading to these countries. When the crossing points are geographical features such as deserts, rivers or seas, the risks of migration become even higher.

Sources: Frontex; Missing Migrants Project; International Organization for Migration; European Union; Riccardo Pravettoni and Francesca Fattori, 'La mort aux frontières', *Le Monde*, 21 November 2022.

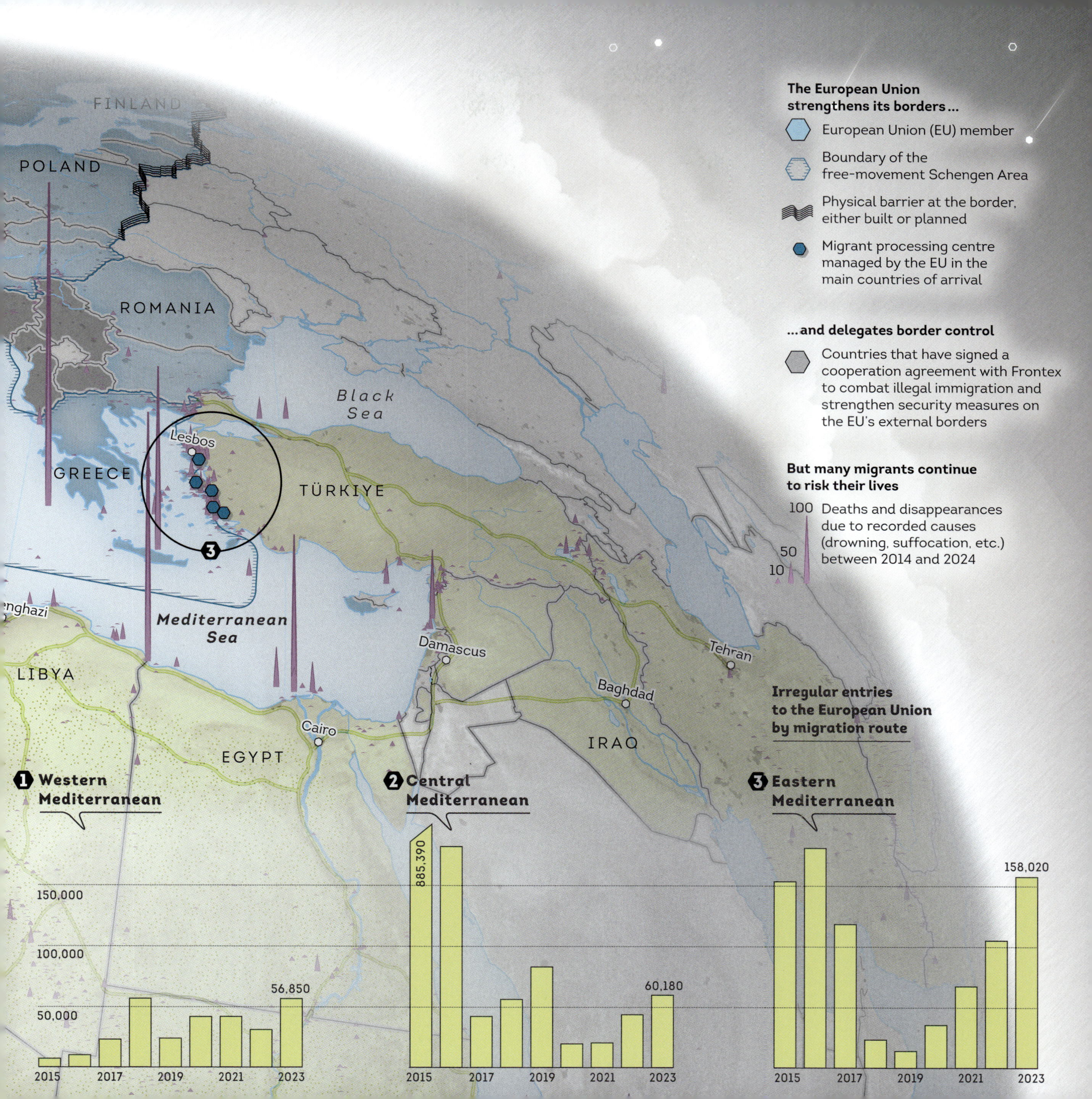

The European Union strengthens its borders ...

European Union (EU) member

Boundary of the free-movement Schengen Area

Physical barrier at the border, either built or planned

Migrant processing centre managed by the EU in the main countries of arrival

... and delegates border control

Countries that have signed a cooperation agreement with Frontex to combat illegal immigration and strengthen security measures on the EU's external borders

But many migrants continue to risk their lives

100 50 10 Deaths and disappearances due to recorded causes (drowning, suffocation, etc.) between 2014 and 2024

FINLAND
POLAND
ROMANIA
Black Sea
Lesbos
GREECE
TÜRKIYE
Benghazi
Mediterranean Sea
LIBYA
Damascus
Tehran
Baghdad
Cairo
EGYPT
IRAQ

Irregular entries to the European Union by migration route

1 Western Mediterranean

2 Central Mediterranean

3 Eastern Mediterranean

150,000
100,000
50,000

56,850
885,390
60,180
158,020

2015 2017 2019 2021 2023
2015 2017 2019 2021 2023
2015 2017 2019 2021 2023

25 THE SCHENGEN AREA

Free movement for 450 million Europeans

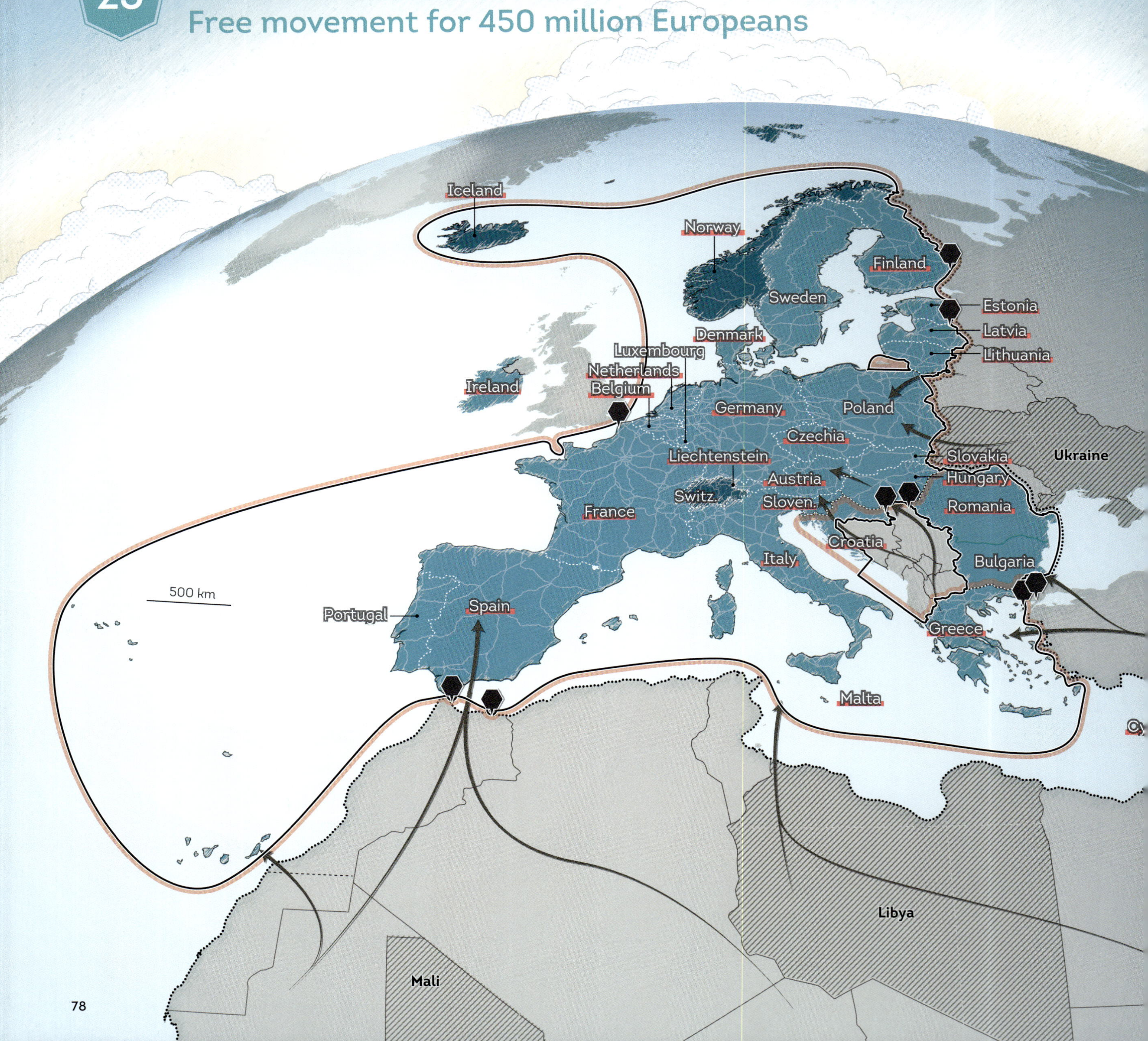

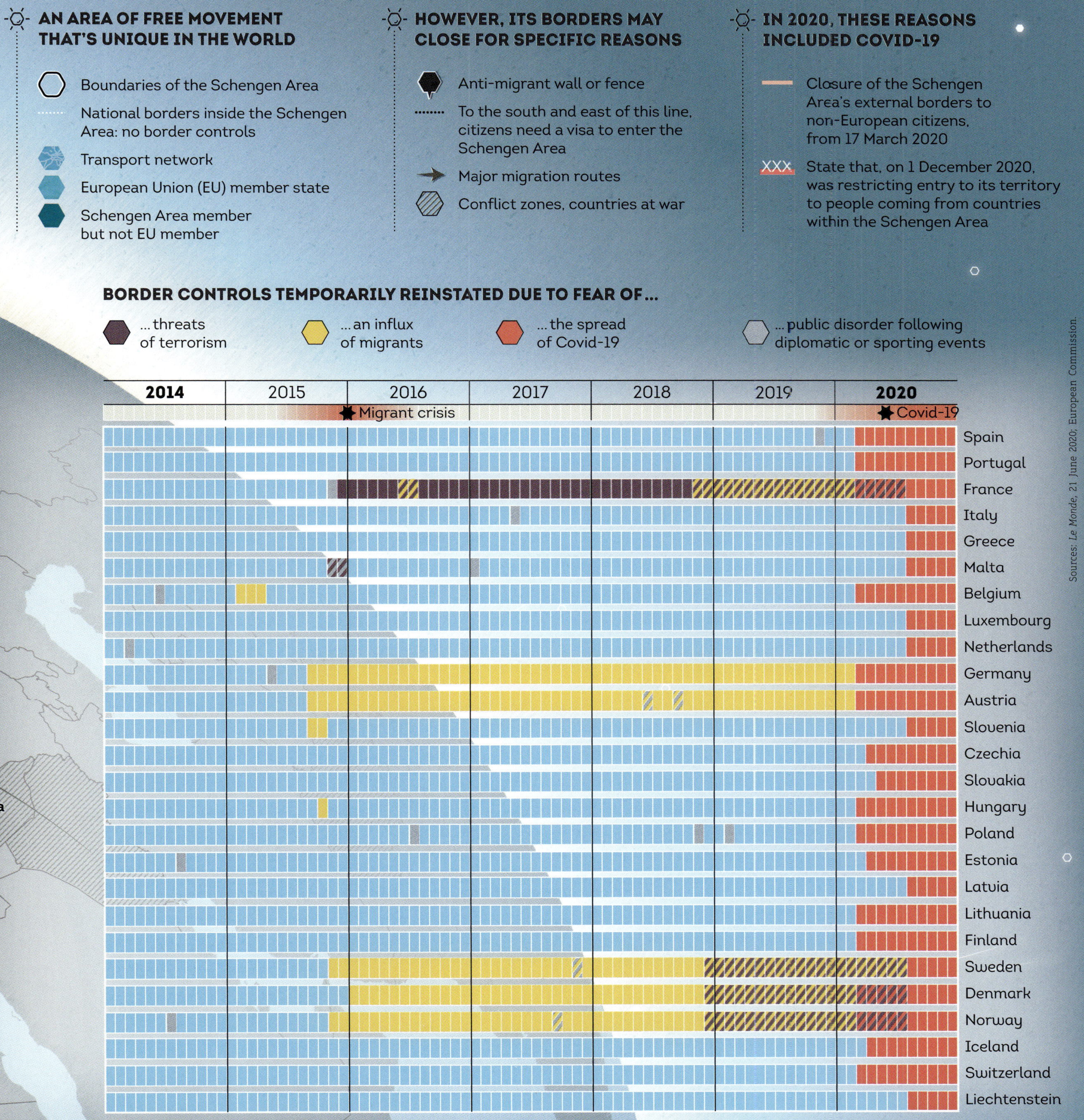

AN AREA OF FREE MOVEMENT THAT'S UNIQUE IN THE WORLD
Boundaries of the Schengen Area
National borders inside the Schengen Area: no border controls
Transport network
European Union (EU) member state
Schengen Area member but not EU member
HOWEVER, ITS BORDERS MAY CLOSE FOR SPECIFIC REASONS
Anti-migrant wall or fence
To the south and east of this line, citizens need a visa to enter the Schengen Area
Major migration routes
Conflict zones, countries at war
IN 2020, THESE REASONS INCLUDED COVID-19
Closure of the Schengen Area's external borders to non-European citizens, from 17 March 2020
XXX State that, on 1 December 2020, was restricting entry to its territory to people coming from countries within the Schengen Area
BORDER CONTROLS TEMPORARILY REINSTATED DUE TO FEAR OF ...
... threats of terrorism
... an influx of migrants
... the spread of Covid-19
... public disorder following diplomatic or sporting events
2014
2015
2016
2017
2018
2019
2020
Migrant crisis
Covid-19
Spain
Portugal
France
Italy
Greece
Malta
Belgium
Luxembourg
Netherlands
Germany
Austria
Slovenia
Czechia
Slovakia
Hungary
Poland
Estonia
Latvia
Lithuania
Finland
Sweden
Denmark
Norway
Iceland
Switzerland
Liechtenstein
Syria
Sources: *Le Monde*, 21 June 2020; European Commission.

PORTUGAL

Faro

SPANISH ENCLAVES

Europe in North Africa

Spain's overseas territories along the Mediterranean coast of North Africa were first occupied under the Roman Empire, then the Reconquista, being used as defence and trading posts. They later became penal colonies, and subsequently grew into urban centres. Until 1969, their borders were open, and today they are still integrated into the legal and illegal economies of their immediate surroundings. As well as the autonomous cities of Ceúta and Melilla, Spain also administers several other *plazas de soberanía* (sovereign territories) off the African coast: the Chafarinas Islands, the Alhucemas Islands, and the rocky tidal island of Peñón de Vélez de la Gomera. A rise in illegal immigration, mainly from sub-Saharan Africa, has led the Spanish government to erect border fences in Ceúta and Melilla.

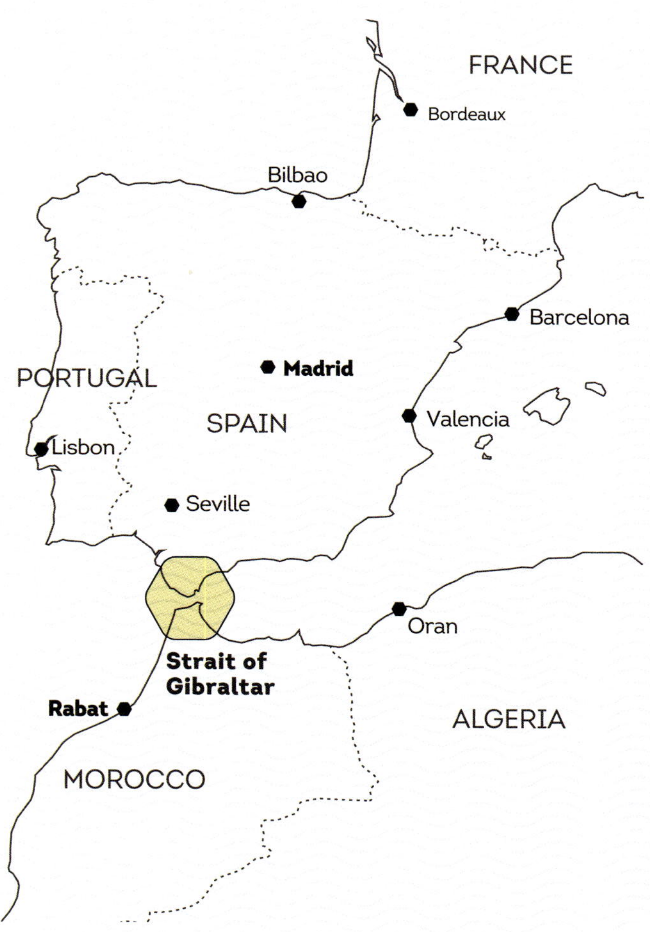

Casablanca

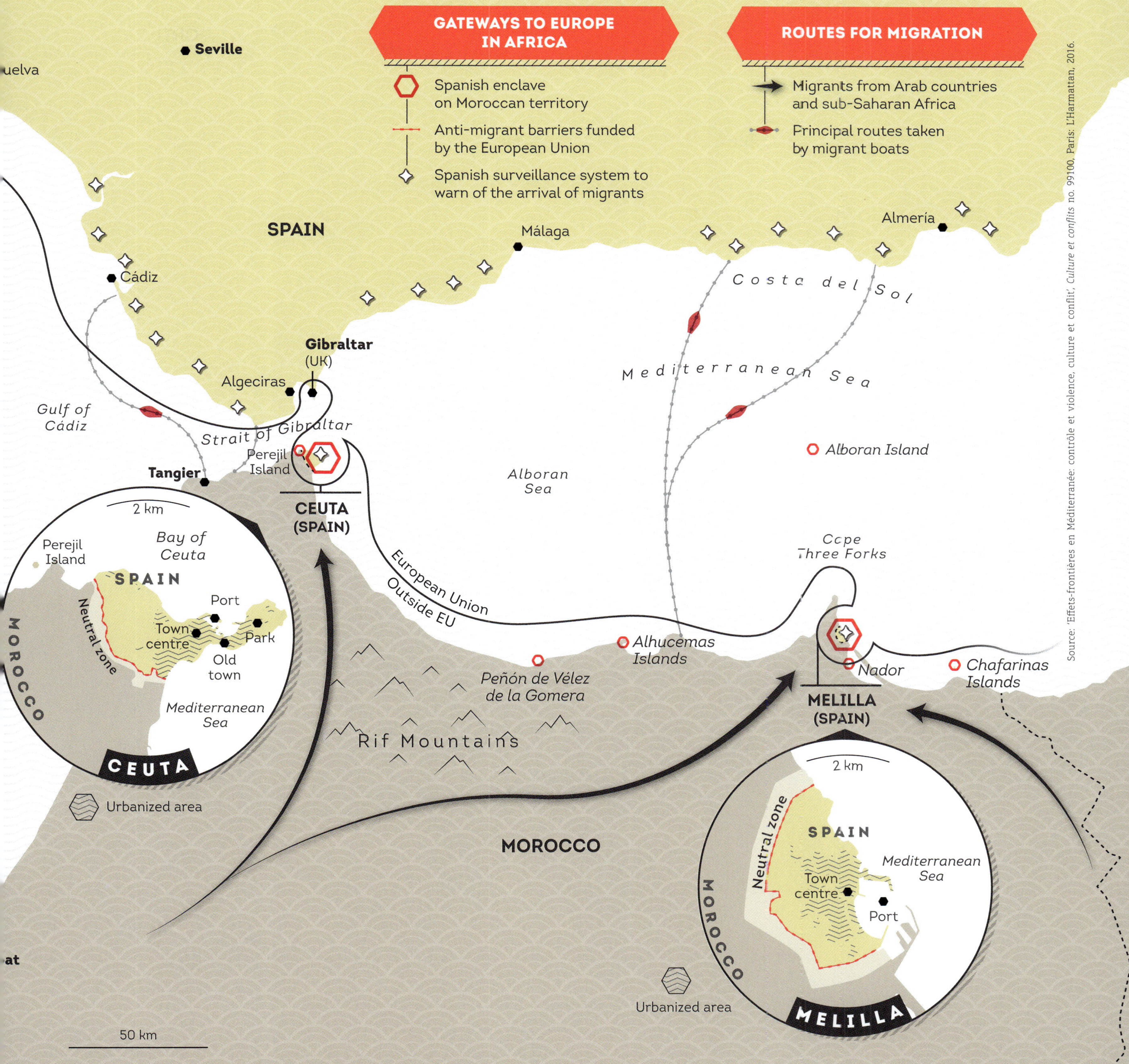

Source: 'Effets-frontières en Méditerranée: contrôle et violence, culture et conflit,' *Culture et conflits* no. 99100, Paris: L'Harmattan, 2016.

27 WORLD PASSPORTS

Keys that don't open every door

RUSSIA

The Joint Security Area in Panmunjom

NORTH KOREA
Joint Security Area
72-Hour Bridge
Military Demarcation Line
Conference Row
Freedom House
Bridge of No Return
Peace House
SOUTH KOREA
2 km

The border between North and South Korea was drawn in 1945 by two young US officers, armed only with a National Geographic map. For convenience's sake, they chose the 38th Parallel, which divided the territory into two roughly equal parts; it so happened that this was also the boundary of Russia and Japan's spheres of influence. Since the armistice in 1953, the Military Demarcation Line, made up of multiple of rows of barbed wire fences, has been surrounded by a demilitarized zone that is 238 km long and 4 km wide. In the centre is the Joint Security Area, where UN soldiers (Americans) stand facing North Korean soldiers. In the middle is the Neutral Nations Supervisory Commission (Switzerland and Sweden on the southern side, Poland and Czechoslovakia on the northern side until they left in 1995). There is also a northern maritime border drawn up by the UN in 1953, but not set out in the armistice agreement. Along with the India–Pakistan border, it is the most heavily militarized border in the world.

Sea of Japan
JAPAN
Tokyo
Kyoto
Hiroshima
Pacific Ocean

Sources: *Hérodote* no. 141, 'Géopolitique de la péninsule coréenne', Paris: La Découverte, 2011/1; Open Street Map.

30 WESTERN SAHARA

A wall in the desert

The Moroccan Western Sahara Wall, also known as the Berm, was built by Morocco a few years after its occupation of Western Sahara in 1975 (the annexation has not been recognized by the UN), to protect the 'useful Sahara'. This barrier, made of sand and stones, is now the longest defensive wall in the world, at more than 2,000 km, although it also serves to control immigration. Two thirds of the Moroccan army, around 100,000 soldiers, are deployed there. In late 2020, for the first time in thirty years, there were clashes between Morocco and the Sahrawi Arab Democratic Republic.

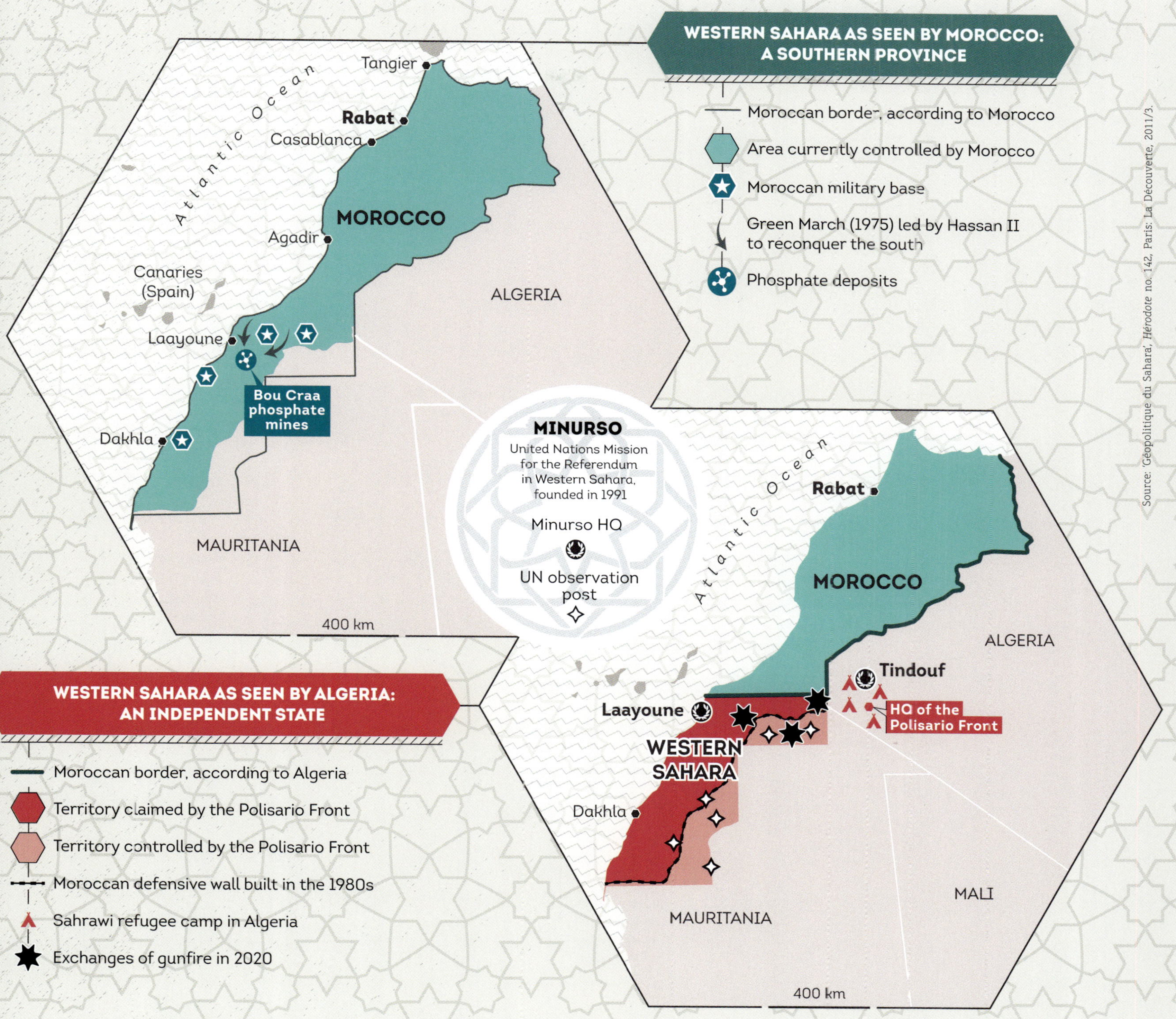

Source: 'Géopolitique du Sahara', *Hérodote* no. 142, Paris: La Découverte, 2011/3.

31

KASHMIR

The highest contested border

Kashmir seen by India

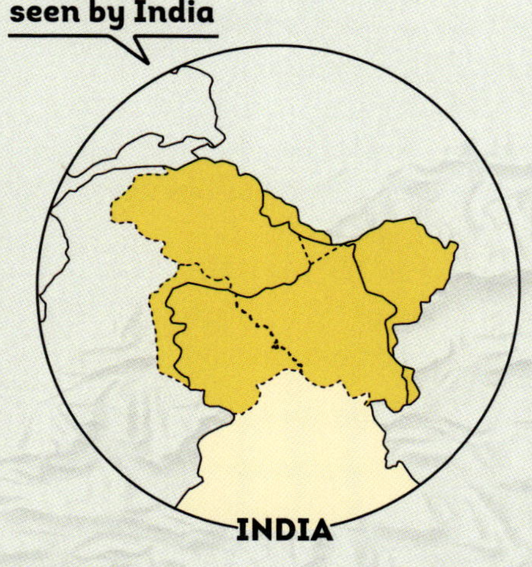

Kashmir seen by Pakistan

Kashmir seen by China

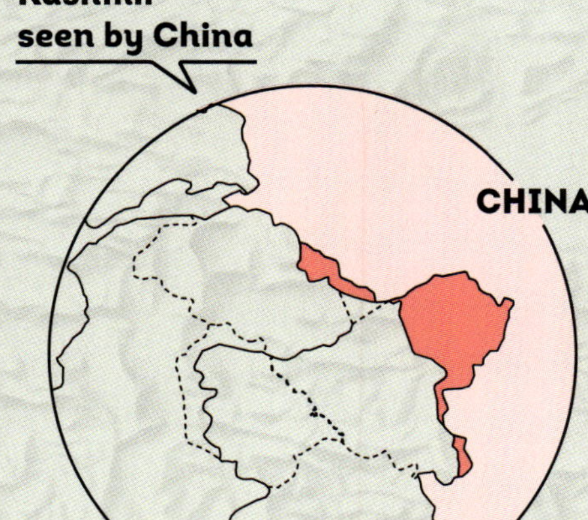

Kashmir is divided between India (60%), Pakistan (30%) and China (10%): an area controlled by India and claimed by Pakistan; an area controlled by Pakistan and that India considers to be occupied; an area (the Shaksgam Valley) that Pakistan ceded to China in 1963 but is claimed by India; and an area (Aksai Chin) controlled by China and also claimed by India. There were serious incidents in summer 2020 between Chinese and Indian troops in the area called the Line of Actual Control. The border with Nepal is also disputed.

Between India and Pakistan, the ceasefire line of 1949 stops when it reaches the Siachen Glacier, at the foot of the Karakoram, at Point NJ9842. Beyond that, the border is not delimited, and there are regular conflicts along the Actual Ground Position Line; it is the highest contested border in the world. The Indian barrier in Kashmir, built at a distance of between 1 and 2 km behind the Line of Control, is 550 km long (with 740 km of the border being contested), running from Point NW605550 to Point NJ9842.

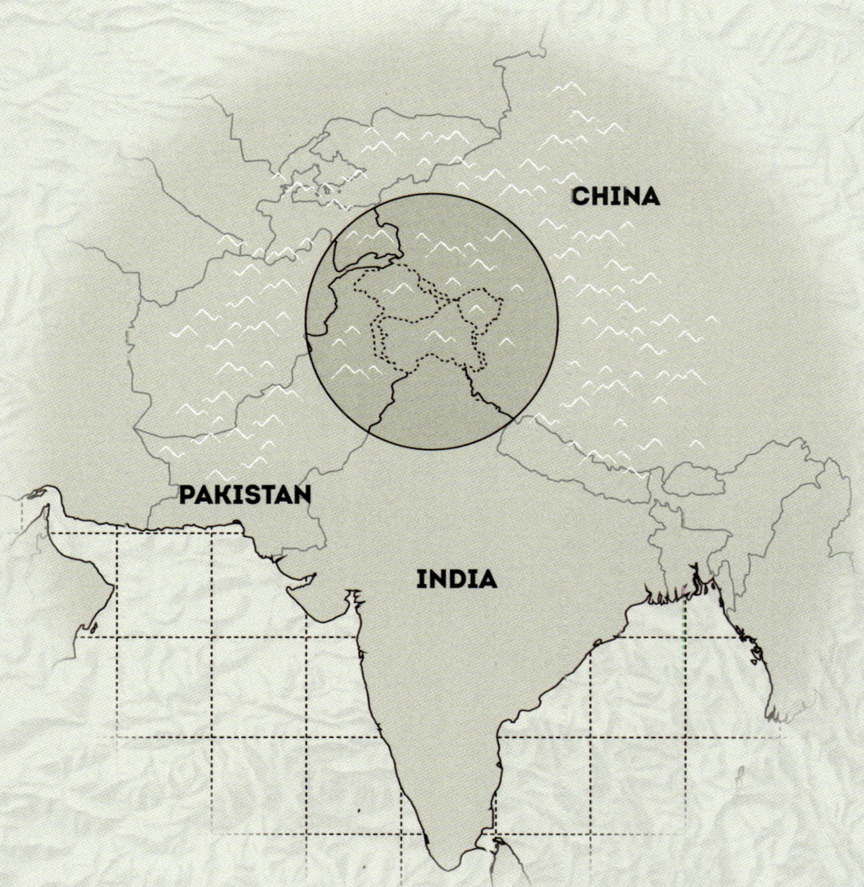

Sources: 'Géopolitique de l'Inde', *Hérodote* no. 173, Paris: La Découverte, 2019/3; *Le Monde*, 17 August 2019.

32 CYPRUS

A divided island

The island of Cyprus has been divided in two since the Turkish invasion in 1974 and the foundation of the 'Turkish Republic of Northern Cyprus' (TRNC), which is only recognized by Türkiye. Today, Nicosia is the only capital city that is divided by a wall. The border is complex: there are British military bases (Akrotiri and Dhekelia, 3% of the island's surface area), enclaves in the middle of these bases, and a buffer zone patrolled by the UN, which also monitors the Greek-Turkish village of Pyla, situated within this zone. The TRNC's territory is theoretically part of the European Union.

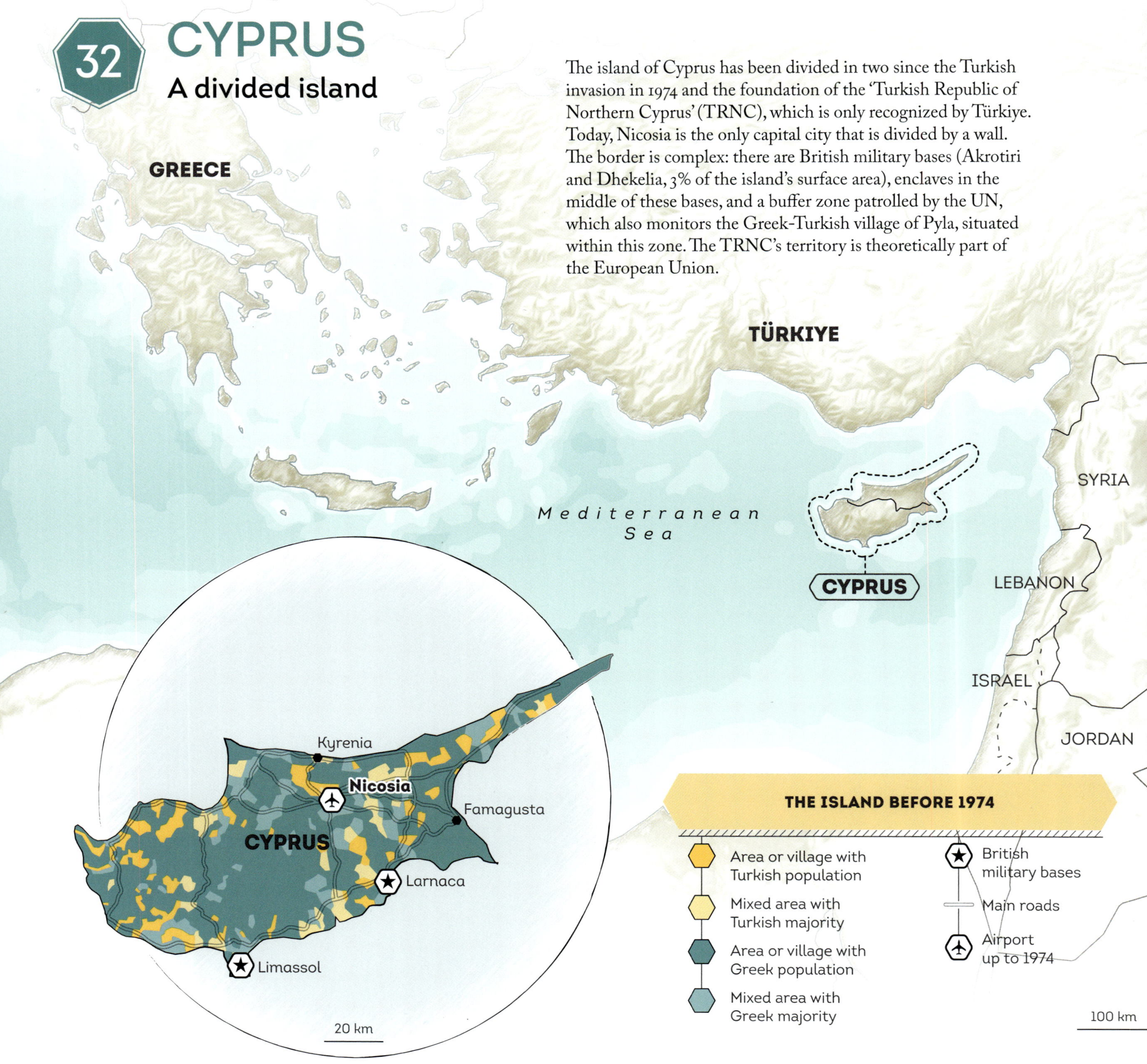

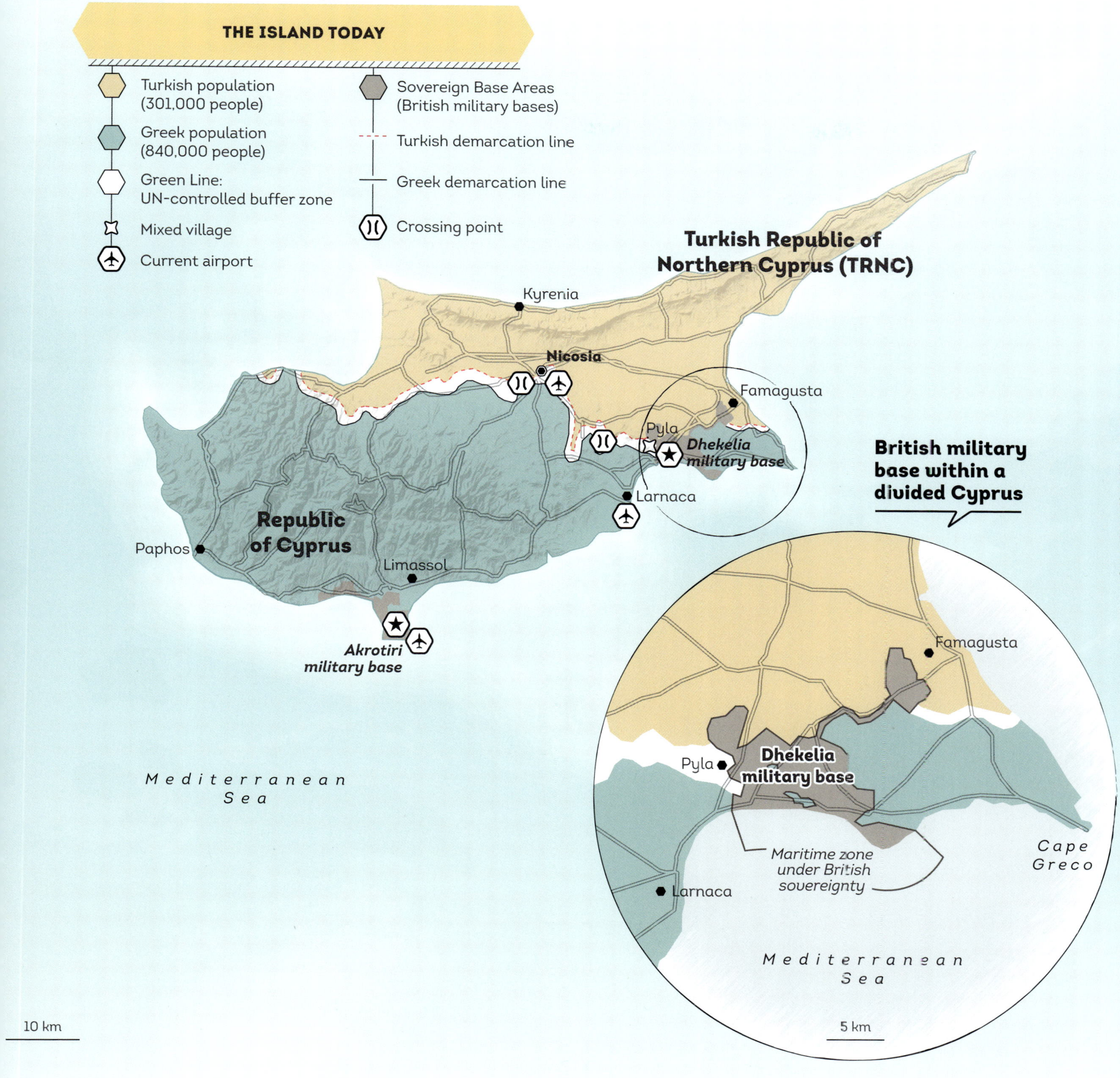

THE ISLAND TODAY

Turkish population (301,000 people)
Greek population (840,000 people)
Green Line: UN-controlled buffer zone
Mixed village
Current airport
Sovereign Base Areas (British military bases)
Turkish demarcation line
Greek demarcation line
Crossing point

Turkish Republic of Northern Cyprus (TRNC)

Kyrenia
Nicosia
Famagusta
Pyla
Dhekelia military base
Larnaca
Republic of Cyprus
Paphos
Limassol
Akrotiri military base

Mediterranean Sea

10 km

British military base within a divided Cyprus

Famagusta
Pyla
Dhekelia military base
Maritime zone under British sovereignty
Cape Greco
Larnaca
Mediterranean Sea

5 km

Source: Pierre Blanc, 'Chypre: un triple enjeu pour la Turquie', *Hérodote* no. 148, Paris: La Découverte 2013/1.

33 WALLS OF THE WORLD

The number of barriers is rising

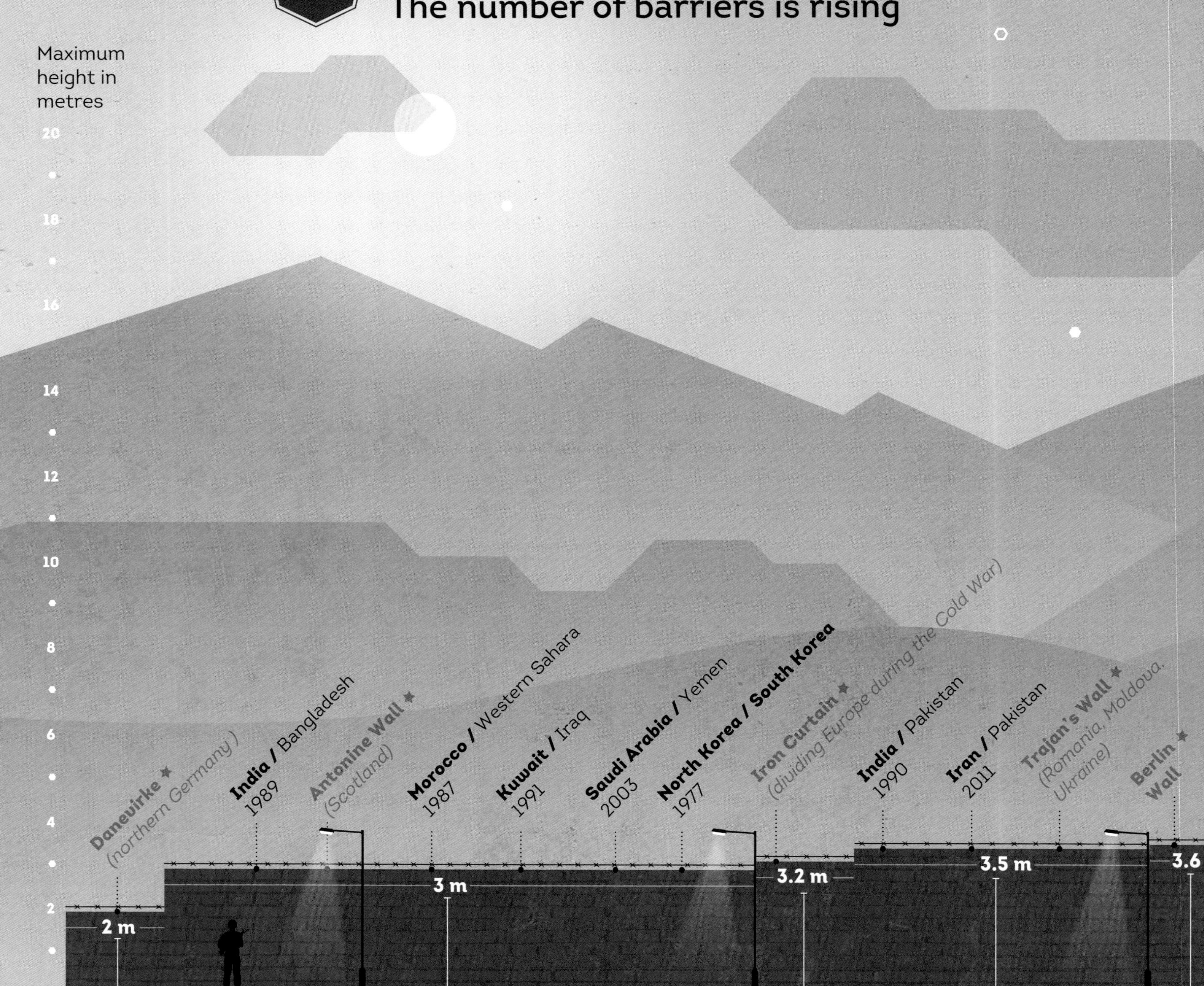

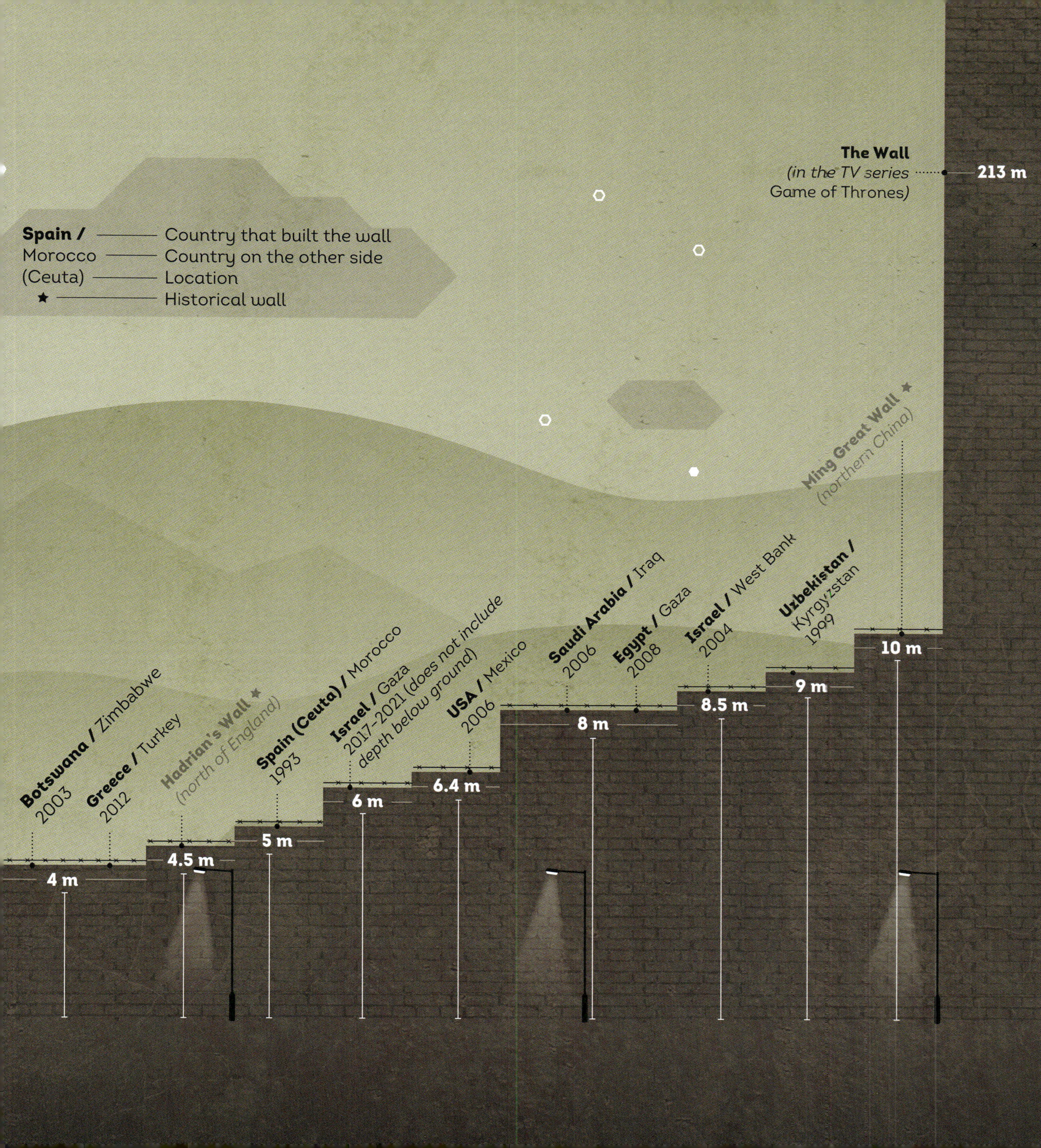

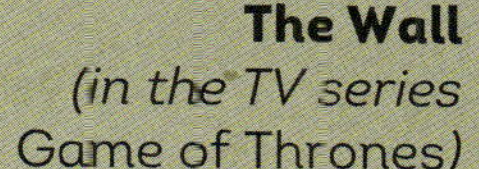

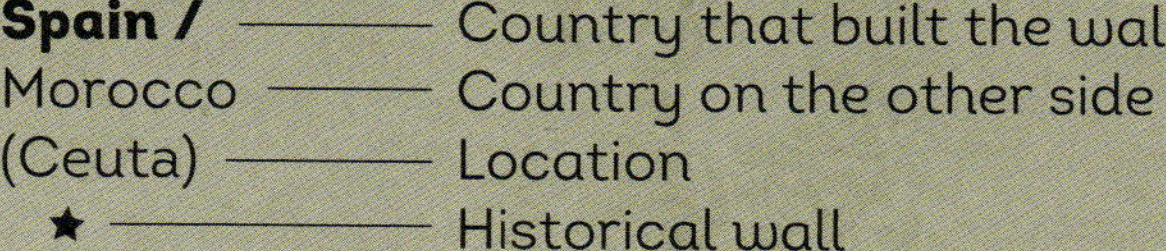

Sources: Alexandra Novosseloff & Frank Neisse, *Des murs entre les hommes*, Paris: La Documentation française, 2015; Alexandra Novosseloff & Frank Neisse, 'Les murs: séparations et traits d'union', *Politique étrangère*, 2010, no. 4.

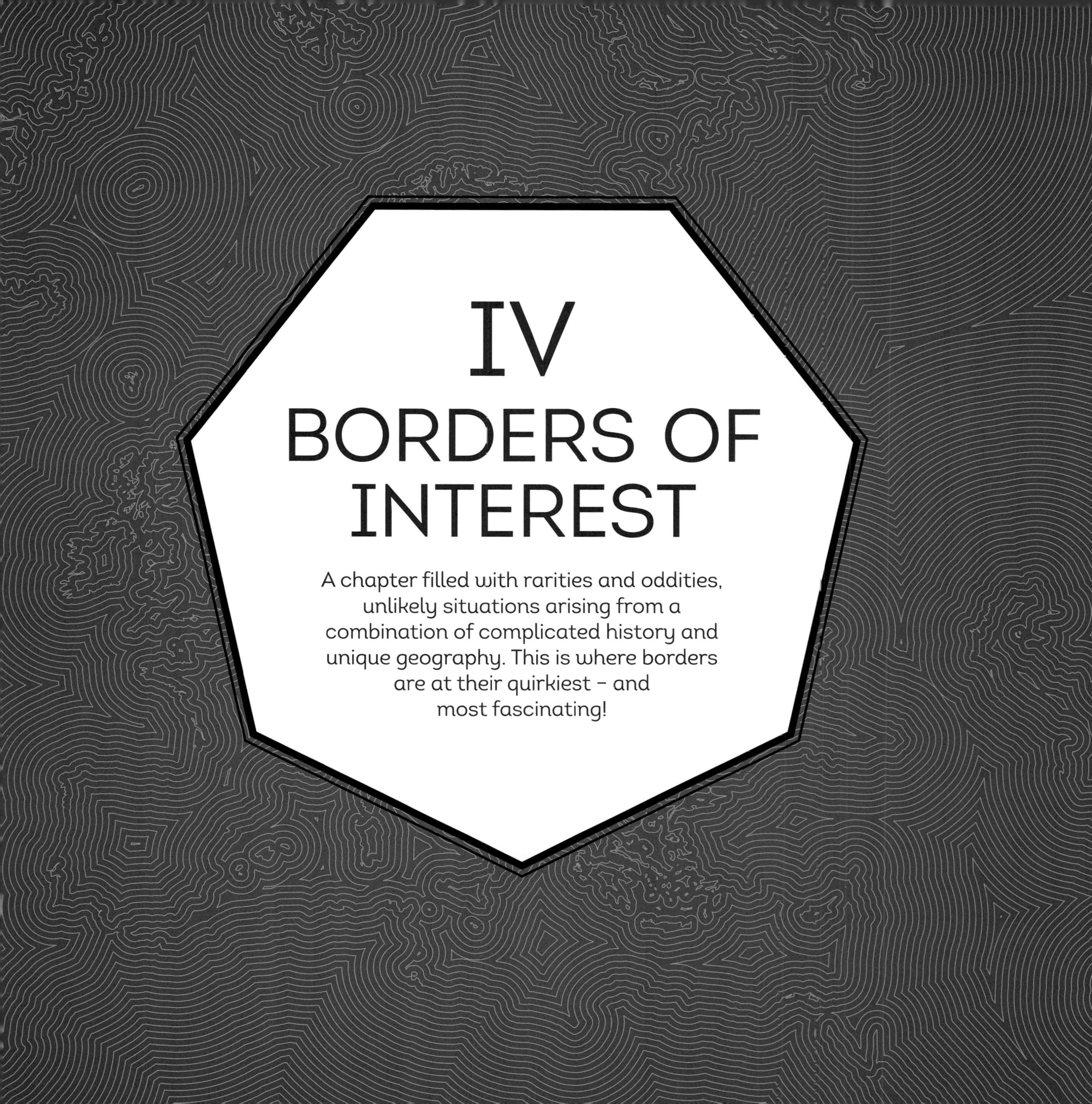

IV
BORDERS OF INTEREST

A chapter filled with rarities and oddities, unlikely situations arising from a combination of complicated history and unique geography. This is where borders are at their quirkiest – and most fascinating!

IV · BORDERS OF INTEREST

From extraterritorial areas to enclaves, from micronations to borders with strange and convoluted shapes, there is a surprising variety of unusual borders in the world. Here are a few of them.

Extraterritorial zones

True extraterritorial zones, where the surrounding state has no sovereignty, are rare. Contrary to popular belief, diplomatic embassies are not fully extraterritorial, although they belong to the state that they represent, and are just as inviolable. The same is true for some religious sites, such as the properties of the Holy See located outside the Vatican City, or the French National Domain in the Holy Land, as well as a number of cemeteries and memorials, including various US and Canadian cemeteries in Europe and elsewhere. The Orthodox monastic community of Mount Athos – where 'no female creature' is allowed to set foot – is considered an autonomous territory of Greece. The territory of Guantánamo Bay in Cuba has been leased to the USA since 1903, the agreement granting complete control and jurisdiction over the area. The small but highly symbolic territory of Tiwintza has belonged to Ecuador since 1995, but is administered by Peru.

So what counts as an extraterritorial area? The headquarters of some international organizations such as the UN are 'international territory'. The seat of the Sovereign Order of Malta, in Rome, is another example, unusual in that the Order is considered an autonomous subject in international law, like a state. What is less clear is the status of the British military bases in Cyprus. Are they a British colony? An exclave of the United Kingdom? A unique easement? The question remains unresolved from a legal standpoint.

Finally, the subtleties of diplomacy can lead to the creation of temporary extraterritorial areas. From 1914 to 1918, Sainte-Adresse in Normandy was the administrative capital of Belgium. From 1999 to 2002, Camp Zeist was made a Scottish enclave in the Netherlands, so that the trial of the Lockerbie bombers could take place on neutral ground, as demanded by Libya. More anecdotally, suite 212 at Claridge's Hotel in London was ceded to Yugoslavia for one day, on 17 July 1945, to allow Queen Alexandra of Yugoslavia to give birth to her son, Prince Alexander, on Yugoslavian territory.

Contrary to popular belief, diplomatic embassies are not extraterritorial: they belong to the state that they represent.

Terra nullius

We live in an era when there is very little space on Earth that can truly be considered *terra nullius*, or 'nobody's land'. Neither – theoretically – are there areas that are outside of any laws, like the 'Debatable Lands' on the border of England and Scotland in the 15th and 16th centuries, where neither English nor Scottish law applied.

The International Court of Justice uses a narrow definition: territories that are inhabited and socially organized, even if occupied, cannot count as *terra nullius*. What's more, *terra nullius* must be unclaimed by any state. According to this definition, neither Antarctica (aside from

Marie Byrd Land) nor the Palestinian Territories can be considered *terra nullius*. Nor can the Greek-Turkish village of Pyla in the UN buffer zone of Cyprus.

The Belgian enclave H22, within Dutch territory, was *terra nullius* until 1996. Today, the territory of Bir Tawil, on the border between Egypt and Sudan, does not belong to anyone: according to the Anglo-Egyptian Treaty of 1899, if either country asserts a right to it, it will lose its rights over the Halaib Triangle, a small border region governed by Egypt since 1995. A few pockets of land administered by Croatia are not claimed by anyone (including Liberland, a small strip of land on the Serbian border, which is a self-proclaimed micronation), as are a few others on the Slovenian border. Finally, Perejil Island is not claimed by either Morocco or Spain, by mutual agreement between the two countries.

A no-man's-land is a deliberately empty space along a frontline or in the area between two border posts. The former no-man's-land that separated Israeli and Jordanian forces from 1948 to 1967 is still marked on maps but in practice it is occupied by Israel.

Enclaves

In a general sense, the word 'enclave' – from the Latin *inclavare*, to lock in – is used to refer to non-island national territories that are separate from the country's main territory. In the strict sense of the word, it only refers to territories that are entirely surrounded by another state; if this is not the case, they are called 'semi-enclaves', 'pene-enclaves' (*pene* meaning 'almost'), or 'functional enclaves'. 'Quasi-enclave' means a territory attached to a state by a single point (e.g. Jungholz in Austria). Often small, but not always (e.g. Alaska), enclaves may be the result of long-ago historical situations (feudalism), a war, colonization, political compromises, omissions or even errors in international treaties. Enclaves can be in the sea: two Malawian islands are entirely surrounded by Mozambican waters; the same is true for Saint Pierre and Miquelon, a French overseas territory surrounded by Canadian waters, and Martín García, an Argentinian island in Uruguayan waters.

Enclaves may also be called 'exclaves' of the territory that they belong to, although in the strict sense of the word, an exclave is a separate territory situated between multiple states, like Nakhchivan.

A number of semi-enclaves have access to the sea. Aside from Alaska (USA), these include Northern Ireland and Gibraltar (UK), Kaliningrad (Russia), Cabinda (Angola), and the Spanish enclaves in Morocco. Gaza may become an exclave of a future Palestinian state. Other semi-enclaves include French Guyana and the French part of the island of Saint-Martin. Lesser known are Point Roberts and Elm Point (USA), Neum (Croatia), Musandam (Oman), Oecusse (East Timor), Temburong (Brunei) and Junagadh (claimed by Pakistan).

Enclaves with no sea access, especially those that are completely surrounded by another state, are more common. They often pose issues concerning the movement of goods and people, which are sometimes partly resolved by neighbouring countries granting permission to pass through their territory without stopping. The most famous one, during the Cold War, was of course West Berlin, between the end of the occupation (1955) and reunification (1990). There were also twelve small

West German enclaves, whose situation was resolved by territory exchanges in 1988. There were about 250 enclaves around the world in 2015, before the issue of Cooch Behar was resolved.

In Europe, most enclaves are simply geographical curiosities: Llívia, a Spanish enclave in French territory (since 1660); Campione d'Italia, an Italian enclave in Switzerland (since 1512); Büsingen am Hochrhein, a German enclave in Switzerland (since 1698); Jungholz, an Austrian enclave in Germany; Međurečje,

It is unlikely that new enclaves will be deliberately created in the future, unless a complex partition process takes place.

a Bosnian enclave in Serbia. There are also four Cypriot enclaves on the British Sovereign Base Areas in Cyprus – which are EU enclaves in non-EU territory, all within an EU member state whose territory is partly occupied. Belgium was granted five German enclaves in the Treaty of Versailles, because of the route of the Vennbahn, a former strategic railway, while the Netherlands has the Belgian enclaves of Baarle. The former Yugoslavia also has its share of enclaves (Bosnian in Serbia, Croatian in Slovenia).

Within the territory of the former Soviet Union, there are a number of enclaves, thae legacy of Soviet policy on minorities, and these are often a source of conflict today, one exception being the small Russian exclave of Sankovo-Medvezhye (surrounded by Belarus). Wars in the Caucasus cut off the Azerbaijani exclave of Nakhchivan, of which Türkiye has been a guarantor since a treaty signed in 1921.

In Central Asia, home to the tangled borders of former Soviet Republics and the Fergana Valley , the situation is no less complicated. It is said that Stalin wanted to create weak republics there: that is true, but their borders also took local needs into account. There are seven recognized enclaves and one de facto enclave. In Kyrgyzstan, there are four Uzbek enclaves (including Sokh, the largest land enclave in the world at 238 square km, whose population is Tajik) and two Tajik enclaves (Kaygarach, Vorukh). In Uzbekistan, there is a Kyrgyz enclave (Barak, de facto enclave) and a Tajik enclave (Sarvan).

In the Middle East, there are very few enclaves, aside from the unique situation of Madha (an Omani territory in the United Arab Emirates), within which is a tiny Emirati counter-enclave (also called a second-order enclave), the village of Nawha.

It is unlikely that new enclaves will be deliberately created in the future, unless a complex partition process takes place. However, this could be a conceivable solution for some of the Israeli settlements in the West Bank, within the context of the creation of an independent Palestinian state.

Other points of interest

Landlocked states

Enclaves should not be confused with landlocked states. This expression may refer to either countries that do not have direct access to the sea, of which there are 45 (including Liechtenstein and Uzbekistan, which are double landlocked), as well as four states or territories that are partly recognized (Kosovo, the West Bank, South Ossetia, Transnistria); or states that are completely surrounded by one other state: the Vatican City, San Marino and Lesotho. Monaco and the Gambia belong to a different category: they are surrounded by another state on their land borders but have access to the sea. The issue of sea access remains critical for some countries, such as Bolivia.

Corridors

Throughout history, some territories – and therefore some borders – have been 'extended' for very specific geopolitical reasons. The Wakhan Corridor in Afghanistan was intended to separate the British and Russian empires (today Tajikistan and Pakistan), thereby allowing London to use Afghanistan as a 'buffer state'. The Caprivi Strip in Namibia (named after Bismarck's successor) was established by the colonial German Empire, in order to facilitate trade by providing access to the Zambezi river.

Tripoints and quadripoints

Places where border points meet are another source of curiosity. 'Tripoints', where the borders of three countries touch, are fairly common.

As for 'quadripoints', these are fairly common for administrative divisions within countries, but do any quadripoints exist between countries? The question remains unresolved. Is there one in southern Africa, the only place on the planet where this could be the case (Namibia–Zambia–Botswana–Zimbabwe)? Or is it actually two tripoints (Namibia–Zambia–Botswana and Zimbabwe–Zambia–Botswana), separated by only 157 metres, as many geographers believe? On a local level, this issue is hotly debated, and has political and commercial implications. Sometimes there has even been gunfire at this border point.

Micronations

What about self-proclaimed micronations? Here are just three of many. Liberland, already mentioned above, is a small territory claimed by both Croatia and Serbia, whose fate has not yet been determined, although it is not *terra nullius* in the legal sense. Another is the Principality of Sealand, which is based on a disused offshore platform in the North Sea. But perhaps the most curious was the Principality of Hutt River, on the west coast of Australia, a simple farm that declared independence, citing an obscure provision in British law.

Condominiums

Examples of joint sovereignty (condominium) are rare. One very unusual case is worth mentioning: France and Spain share sovereignty of Conference Island (also known as Pheasant Island, or the Island of Diplomats) in the Bidasoa River in the Basque region, with the two countries' naval authorities each assuming the title of viceroy of the island for alternating periods of six months.

34 GUANTÁNAMO BAY

A US enclave in Cuba

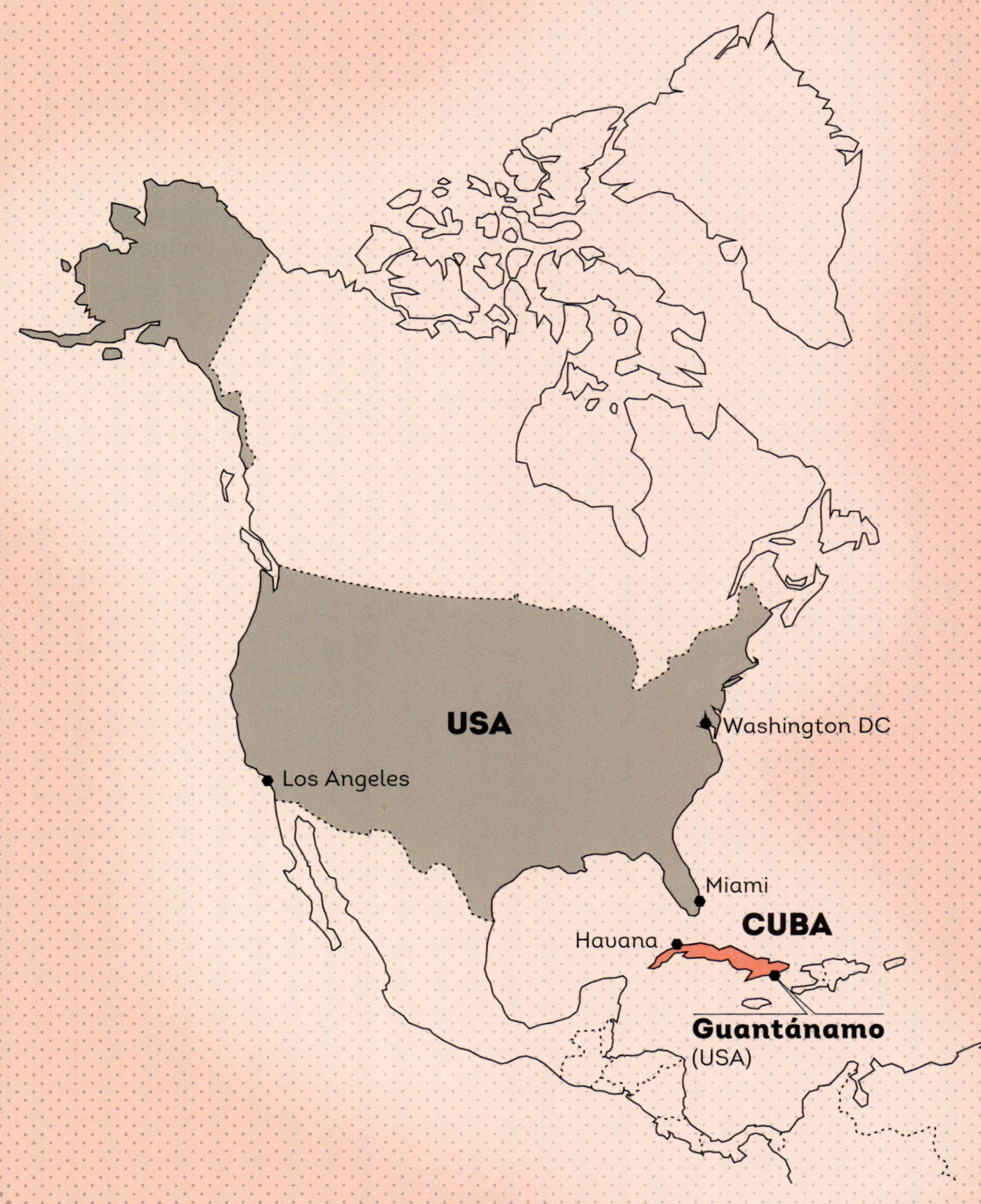

A U.S. NAVAL BASE WITH A PERPETUAL LEASE SINCE 1903

Perimeter fence 28 km long, protected by Marines on the US side and mines on the Cuban side

Guantánamo Bay deep and wide enough to allow US Navy ships to enter and dock

Military HQ

Accommodation

Hospital

THE BASE IS NOW USED BY THE U.S. AS AN OVERSEAS DETAINMENT CAMP

Camp X-Ray, with open-air cages, was opened in 1994 to hold prisoners linked to the recent coup in Haiti. In January 2002 it became a temporary detainment camp for suspected terrorists

Camp Delta was built to replace Camp X-Ray. It contains 600 cells

Building AV-624 is a former barracks refurbished as accommodation for lawyers, journalists and other visitors

Court

The US enclave of Guantánamo Bay (116 square km) has been leased from Cuba since 1903, with no end date. It is home to the oldest overseas US naval base. For security reasons (due to its isolation) and legal reasons, the Guantánamo camp was chosen in 2002, during the 'War on Terror', as the detainment centre for 'unlawful enemy combatants' from 35 different countries (almost 800 in total).

The United States has exclusive jurisdiction but the territory comes under Cuban sovereignty, which allowed the Bush administration to claim that detainees in the camp did not have the same rights (*habeas corpus*) as they would if they had been held on US soil. As of January 2025, there were still 15 prisoners in the detention centre, but the second Trump administration has plans to send more detainees there.

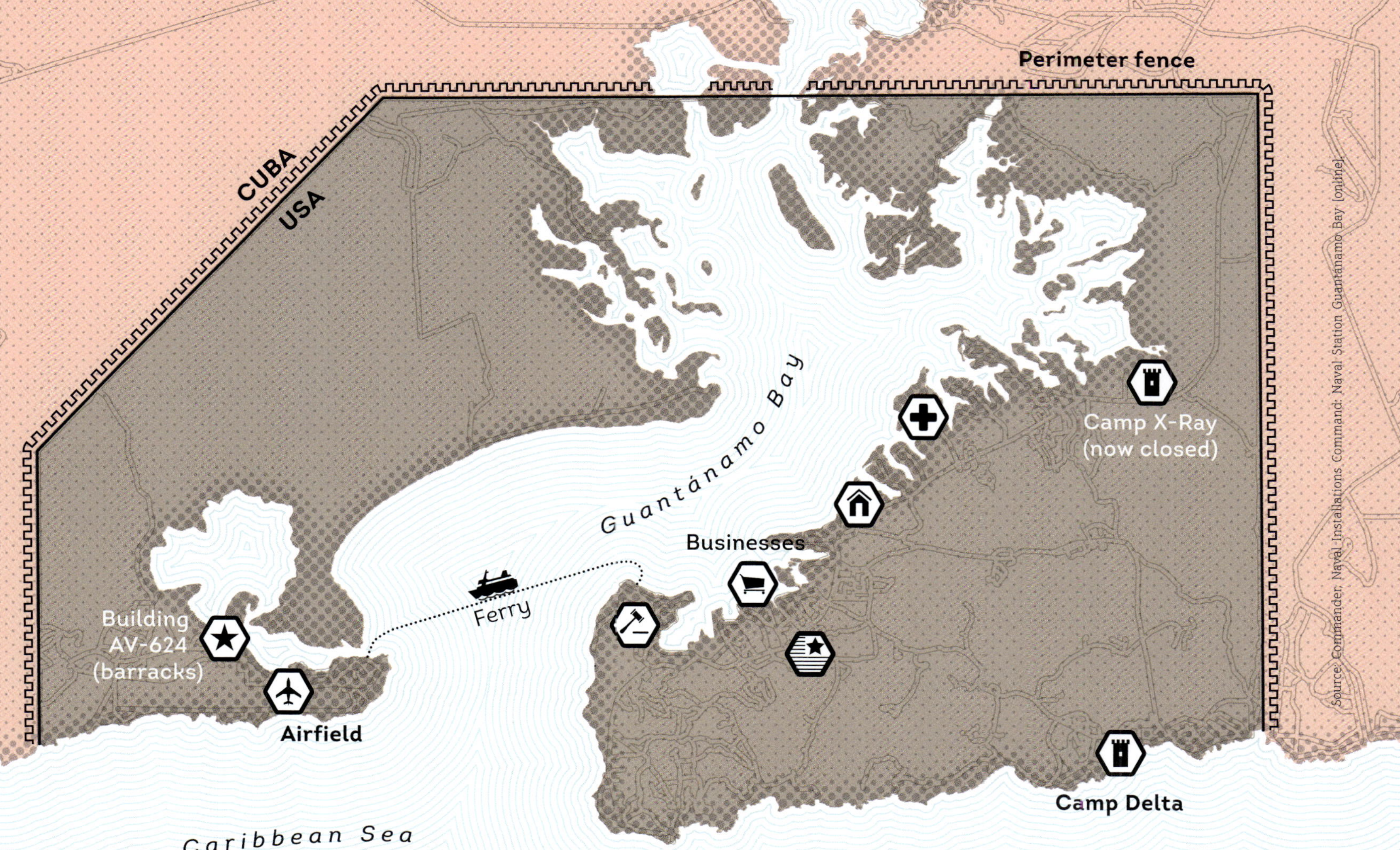

Source: Commander, Naval Installations Command: Naval Station Guantánamo Bay [online]

THE ENCLAVES OF COOCH BEHAR

Perhaps the most complex border in the world

The region of Cooch Behar was home to 192 tiny enclaves – called *chhitmahals* ('crumbs of land'), measuring between 1 hectare and 20 square km – that trace their origins back to the Mughal Empire and a treaty signed in 1713. They served as safe havens during the religious conflicts of 1965–71. But after the closing of the border, which stretches for 3,200 km, life for their 50,000 inhabitants became even more difficult. The enclaves contained three counter-enclaves and one counter-counter-enclave, that of Dahala Khagrabari (0.69 hectares): a tiny piece of Indian territory, within Bangladeshi land, which in turn was surrounded by Indian territory, all inside Bangladesh. The two countries resolved the issue almost completely in May 2015, signing an agreement that laid out an exchange of 162 first-order enclaves (111 Indian, 51 Bangladeshi), which opened up the three counter-enclaves and the single counter-counter-enclave. India renounced its claim to 40 square km of land, whose inhabitants were allowed to choose which nation to belong to. The largest Bangladeshi enclave, Dahagram-Angarpota, which has been connected to its mother country by a land corridor since 2011, was not part of the agreement. The matter of enclaves is a sensitive one for India, due to concerns over terrorism, immigration and trafficking. The border, monitored by 22,000 guards and 70,000 soldiers, runs alongside the Siliguri Corridor, also known as the Chicken's Neck. This strip of land, which separates Nepal from Bangladesh, is a vulnerable spot for India. It was expanded in 1975 when the kingdom of Sikkim was integrated into India, forming a buffer zone between Nepal and Bhutan.

Bangladesh,
A FORMER EXCLAVE OF PAKISTAN

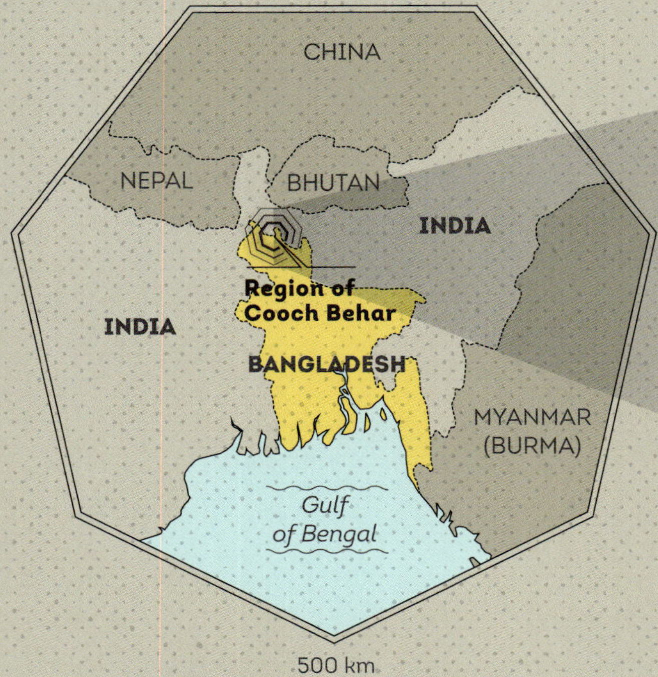

Along the border:
BANGLADESHI ENCLAVES IN INDIA

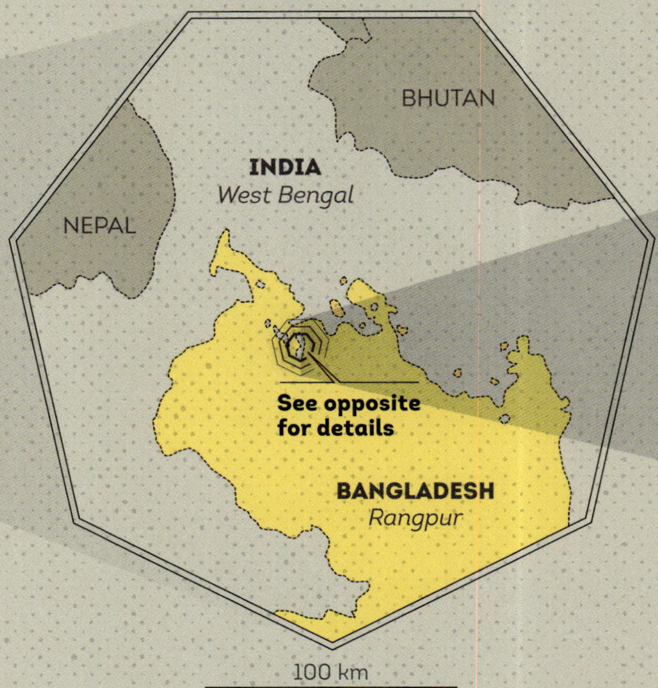

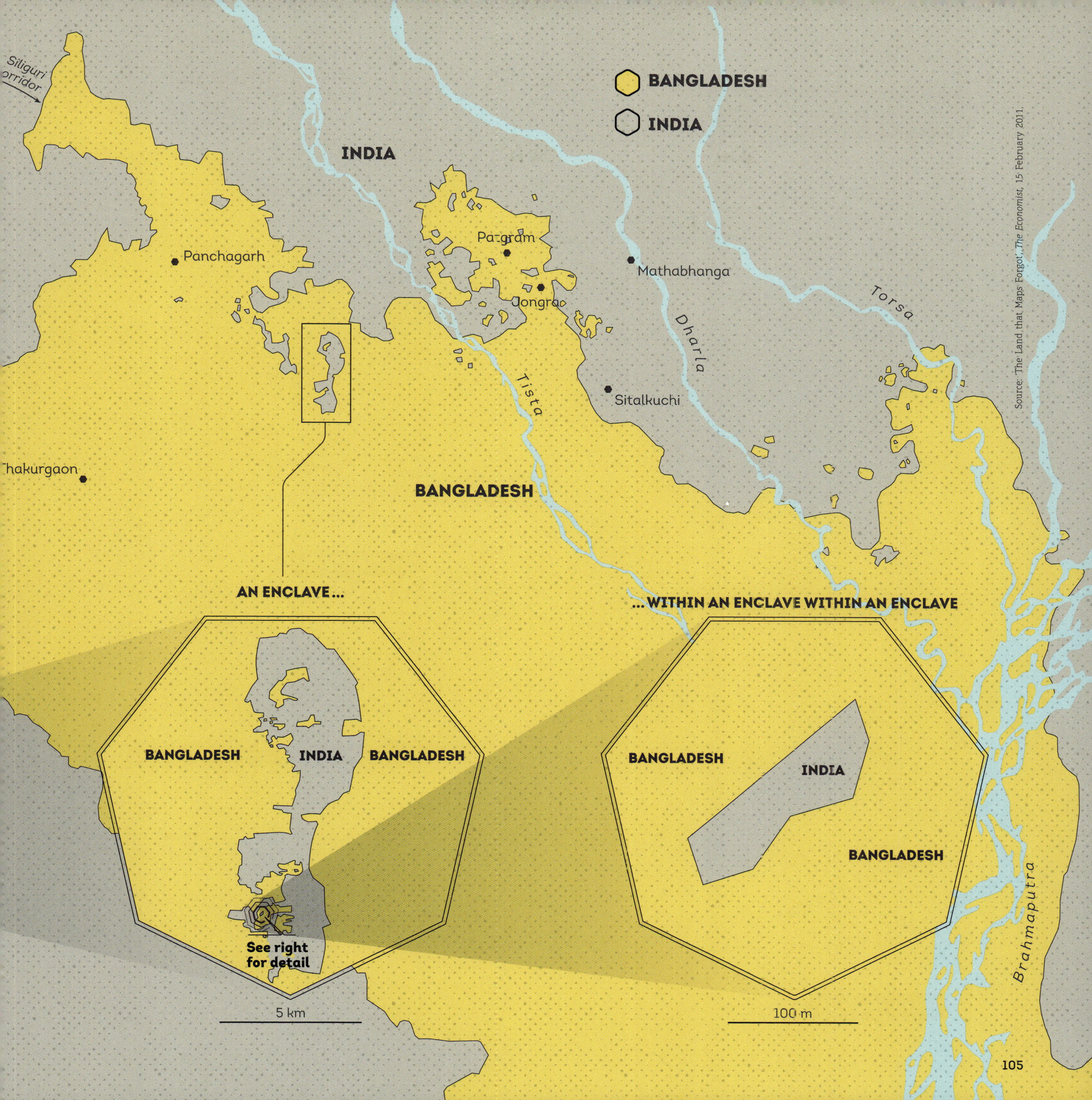

Source: 'The Land that Maps Forgot', *The Economist*, 15 February 2011.

BAARLE-HERTOG & BAARLE-NASSAU
OVERLAPPING TOWNS
Belgian enclave in Dutch territory
Belgian institutions
Dutch institutions
N639
To Breda (North Brabant)
N260
To Tilburg (North Brabant)
Het Goordonk
Onze-Lieve-Vrouw van Bijstand Church
Baarle-Nassau Town Hall
St Remigius Church
Baarle-Hertog Town Hall
BELGIUM
NETHERLANDS
Baarle-Hertog
Town of Baerle-Nassau (NETHERLANDS)
500 m
NETHERLANDS
BELGIUM
Baarle-Hertog
NETHERLANDS
BELGIUM
To Turnhout (Flemish region)
N260

THE BORDERS OF BAARLE

Where Belgium and the Netherlands overlap

Since the Middle Ages, Belgium has possessed 22 small territories in the Netherlands (the commune of Baarle-Hertog), within which are eight Dutch counter-enclaves (the commune of Baarle-Nassau). They are so closely interconnected that the nation to which a house belongs may depend on where its front door is located. One house even belongs to both countries, because its front door straddles the border. The citizens live in peaceful coexistence: on some border roads, bars and restaurants move their tables around – from one country to the other – at closing time, if the neighbouring country's licensing hours are later.

Amsterdam
The Hague
NETHERLANDS
BAARLE-HERTOG
Antwerp
Brussels
BELGIUM

Sources: Official websites of Baarle-Hertog and Baarle-Nassau; background: Open Street Map.

37 UNUSUAL BORDERS

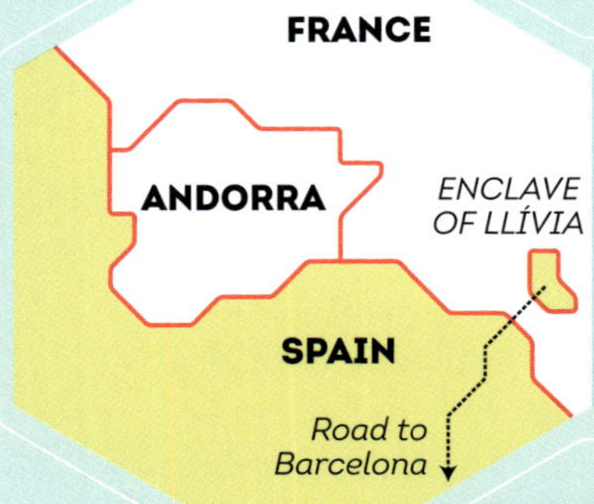

This Spanish village in French territory, whose long-debated status was settled in 1866, is connected to its mother country by a neutral road, with no border controls.

THE ENCLAVE OF LLÍVIA
A LITTLE PIECE OF SPAIN IN THE FRENCH PYRENEES

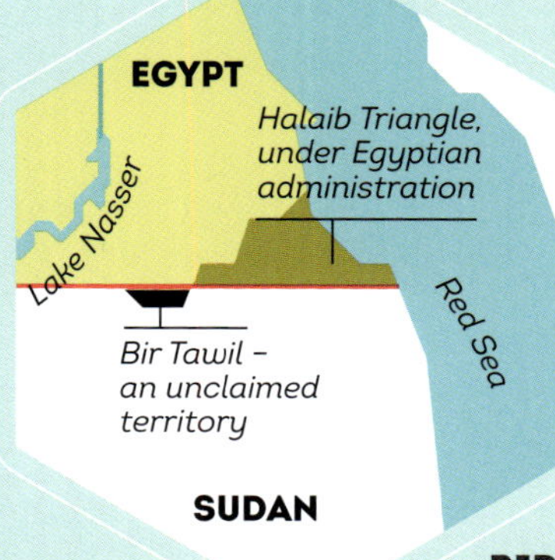

The territory of Bir Tawil is unclaimed. Because of an Anglo-Egyptian treaty signed in 1899, if either Egypt or Sudan asserted a claim to this piece of land, it would lose its rights over the Halaib Triangle, a neighbouring border region administered by Egypt but claimed by both nations.

BIR TAWIL
UNCLAIMED LAND

Established by a treaty signed by Sweden and the Russian Empire in 1809, the border that runs across this small island is both very short and very complicated. In 1985, it was redrawn so that the island's lighthouse would be on Finnish territory, but Sweden would not lose any land.

MÄRKET ISLAND
THE SMALLEST ISLAND DIVIDED BY A BORDER

Because of the Treaty of Versailles (1919), this German former railway line lies within Belgian territory, with the land and most of the stations along it belonging to Belgium. It is now a cycle path.

THE VENNBAHN
A BELGIAN RAILWAY ON GERMAN TERRITORY

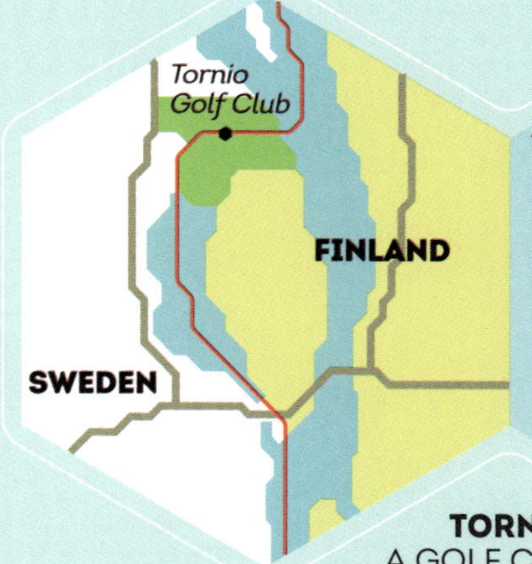

This golf course stretches across Finnish and Swedish territory. One of the holes straddles the border, which can make tee-off times a little complicated, as the two countries belong to different time zones. This means you can hit the ball in Sweden and it will land in Finland an hour later!

TORNIO
A GOLF COURSE THAT CROSSES A BORDER

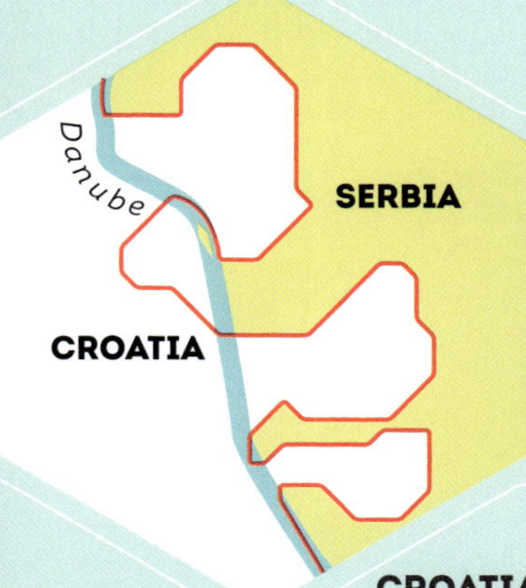

Serbia considers the border to run along the current thalweg, while Croatia says it is defined by the boundaries of the neighbouring municipalities, which are based on the thalweg of the river as it was in the 19th century.

CROATIA–SERBIA BORDER
THE FORMER PATH OF THE DANUBE FORMS THE CURRENT BORDER

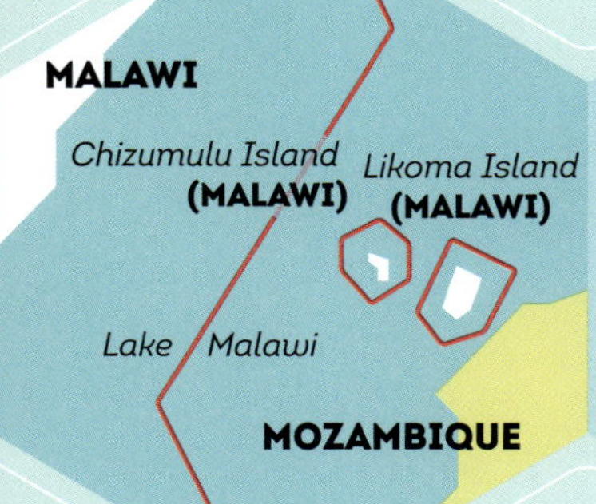

These two Malawian exclaves were given to the country because there were British missionaries on one of the two islands. At the time, Malawi was a British protectorate, while Mozambique was a Portuguese colony.

ISLANDS OF LAKE MALAWI
A LEGACY OF THE COLONIAL ERA

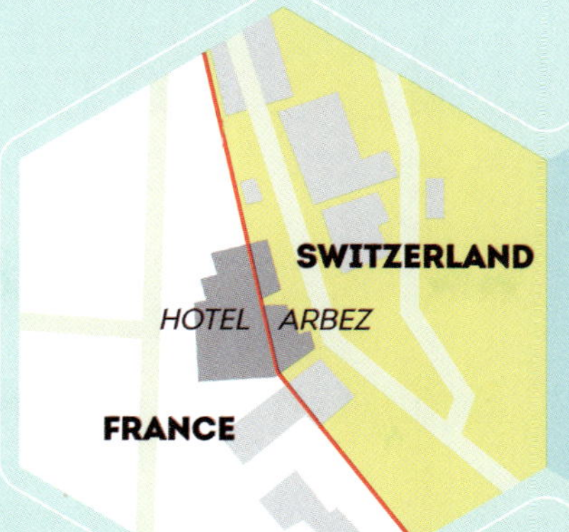

The Hotel Arbez was built in 1895 by a property developer who wanted to exploit the ambiguous status of a piece of land on the French-Swiss border, due to a planned boundary change. The border passes through the hotel's dining room, staircase and some of the bedrooms.

A HOTEL IN TWO COUNTRIES
ON THE FRENCH-SWISS BORDER

Since 1866, this island has been a condominium with alternating sovereignty: the island changes hands every six months between France and Spain. It is also known as Conference Island because the two nations signed the Treaty of the Pyrenees there in 1659, establishing peace between them.

PHEASANT ISLAND
AN ISLAND WITH SHARED SOVEREIGNTY

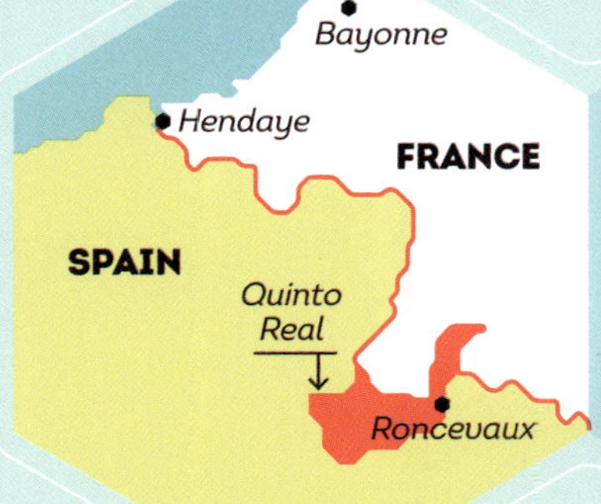

Quinto Real (Kintoa in Basque) is a tiny Spanish territory under French administration. Its inhabitants are French citizens but taxes are split between France and Spain. Its name comes the fact that historically, the local barons had the right to one-fifth (*quinto* in Spanish) of any cargo passing through the area.

QUINTO REAL
SPANISH LAND UNDER FRENCH ADMINISTRATION

Multiple sources.

38 MORE UNUSUAL BORDERS

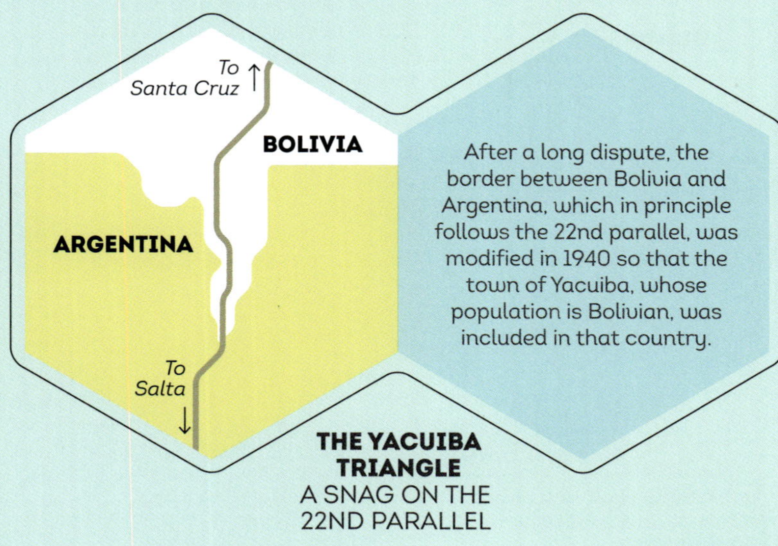

After a long dispute, the border between Bolivia and Argentina, which in principle follows the 22nd parallel, was modified in 1940 so that the town of Yacuiba, whose population is Bolivian, was included in that country.

THE YACUIBA TRIANGLE
A SNAG ON THE 22ND PARALLEL

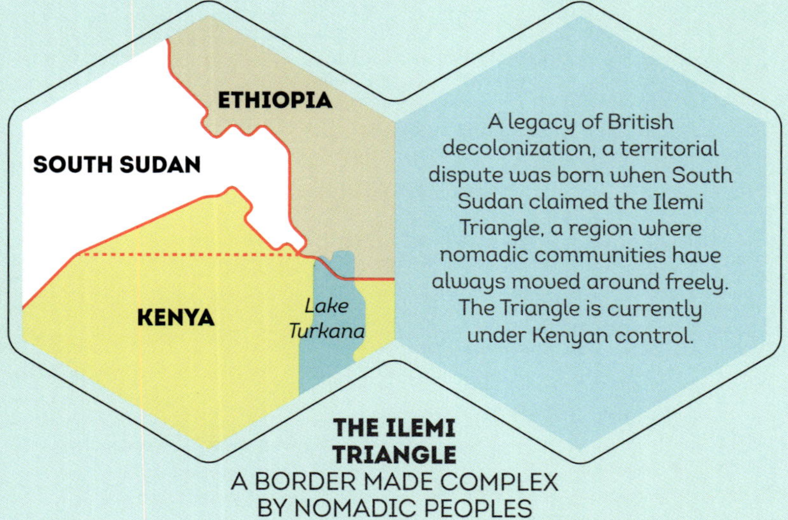

A legacy of British decolonization, a territorial dispute was born when South Sudan claimed the Ilemi Triangle, a region where nomadic communities have always moved around freely. The Triangle is currently under Kenyan control.

THE ILEMI TRIANGLE
A BORDER MADE COMPLEX BY NOMADIC PEOPLES

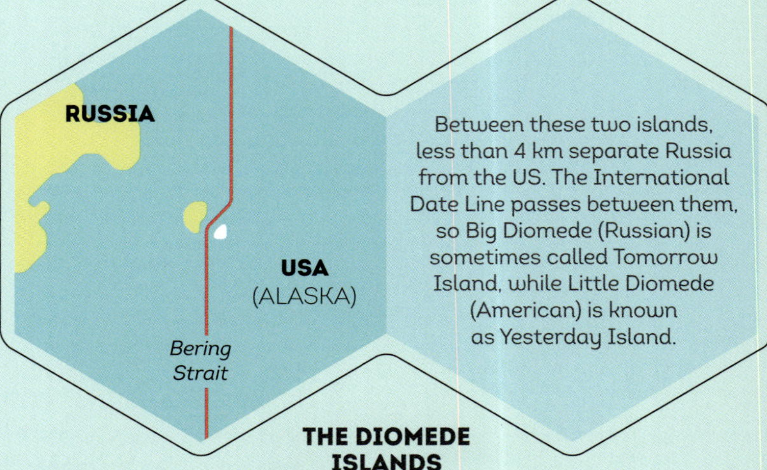

Between these two islands, less than 4 km separate Russia from the US. The International Date Line passes between them, so Big Diomede (Russian) is sometimes called Tomorrow Island, while Little Diomede (American) is known as Yesterday Island.

THE DIOMEDE ISLANDS
SEPARATED BY THE INTERNATIONAL DATE LINE

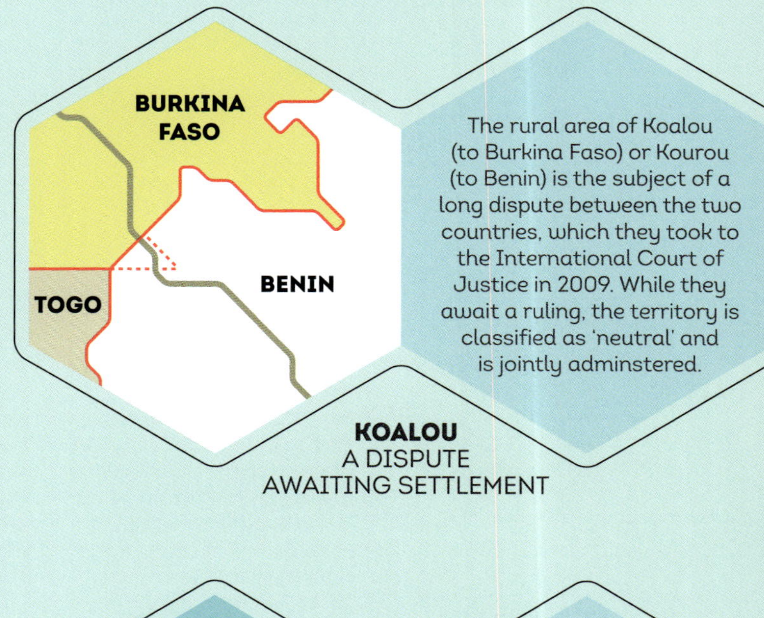

The rural area of Koalou (to Burkina Faso) or Kourou (to Benin) is the subject of a long dispute between the two countries, which they took to the International Court of Justice in 2009. While they await a ruling, the territory is classified as 'neutral' and is jointly adminstered.

KOALOU
A DISPUTE AWAITING SETTLEMENT

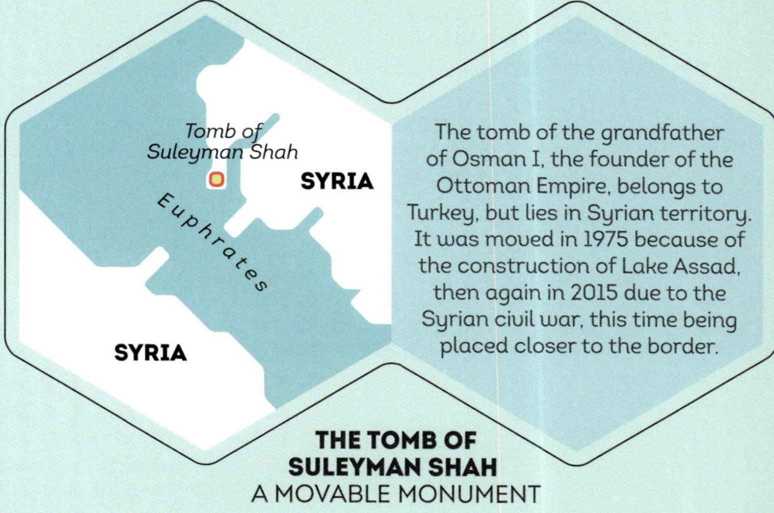

The tomb of the grandfather of Osman I, the founder of the Ottoman Empire, belongs to Turkey, but lies in Syrian territory. It was moved in 1975 because of the construction of Lake Assad, then again in 2015 due to the Syrian civil war, this time being placed closer to the border.

THE TOMB OF SULEYMAN SHAH
A MOVABLE MONUMENT

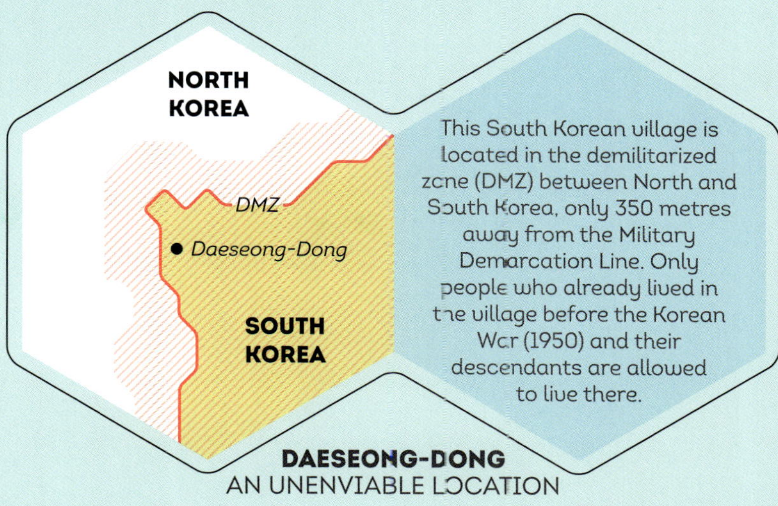

This South Korean village is located in the demilitarized zone (DMZ) between North and South Korea, only 350 metres away from the Military Demarcation Line. Only people who already lived in the village before the Korean War (1950) and their descendants are allowed to live there.

DAESEONG-DONG
AN UNENVIABLE LOCATION

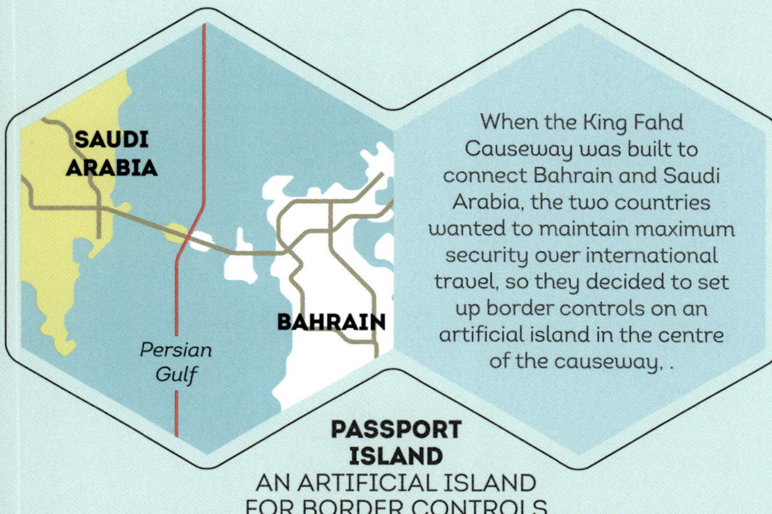

When the King Fahd Causeway was built to connect Bahrain and Saudi Arabia, the two countries wanted to maintain maximum security over international travel, so they decided to set up border controls on an artificial island in the centre of the causeway, .

PASSPORT ISLAND
AN ARTIFICIAL ISLAND FOR BORDER CONTROLS

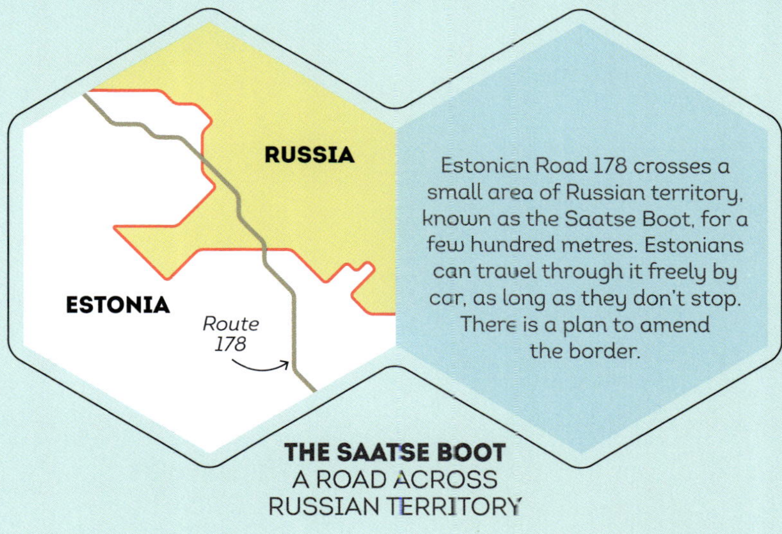

Estonian Road 178 crosses a small area of Russian territory, known as the Saatse Boot, for a few hundred metres. Estonians can travel through it freely by car, as long as they don't stop. There is a plan to amend the border.

THE SAATSE BOOT
A ROAD ACROSS RUSSIAN TERRITORY

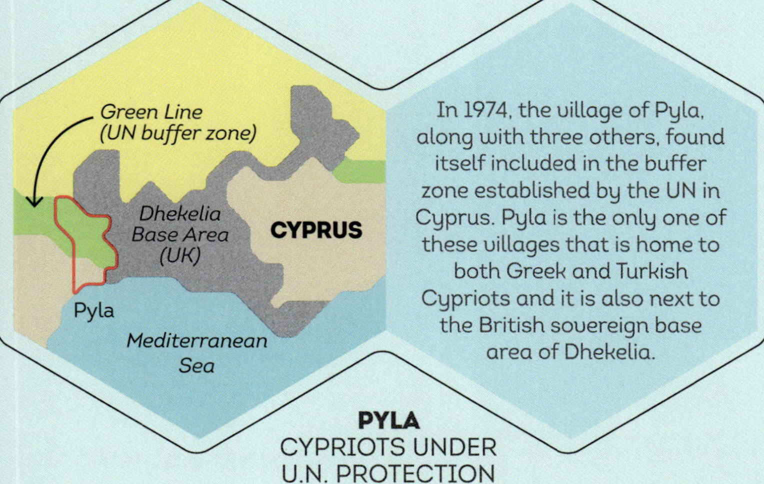

In 1974, the village of Pyla, along with three others, found itself included in the buffer zone established by the UN in Cyprus. Pyla is the only one of these villages that is home to both Greek and Turkish Cypriots and it is also next to the British sovereign base area of Dhekelia.

PYLA
CYPRIOTS UNDER U.N. PROTECTION

The district of Brčko lies in Bosnia and Herzegovina, on the Croatian border. It is considered part of both political entities that make up Bosnia and Herzegovina: the Federation of Bosnia and Herzegovina and the Bosnian Serb Republic (Republika Srpska). In fact, it divides the north and south regions of the Serb Republic.

BRČKO
A LEGACY OF THE YUGOSLAV WARS

Source: Bruno Tertrais.

39 RECORD-BREAKING BORDERS

THE **SHORTEST**
BORDER
80 metres

THE **OLDEST**
BORDER
1278

THE **LONGEST**
BORDER
8,991 km

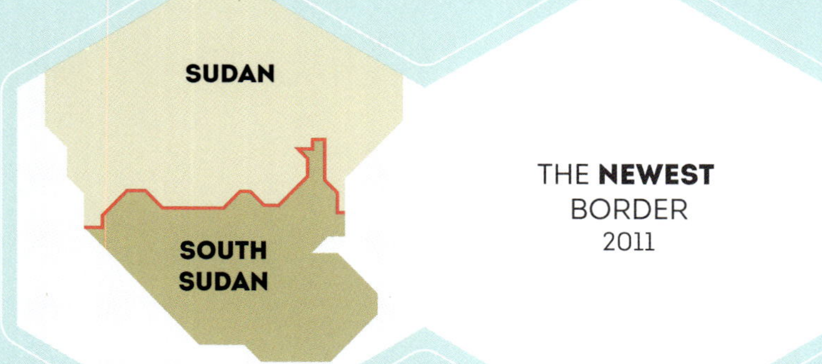

THE **NEWEST**
BORDER
2011

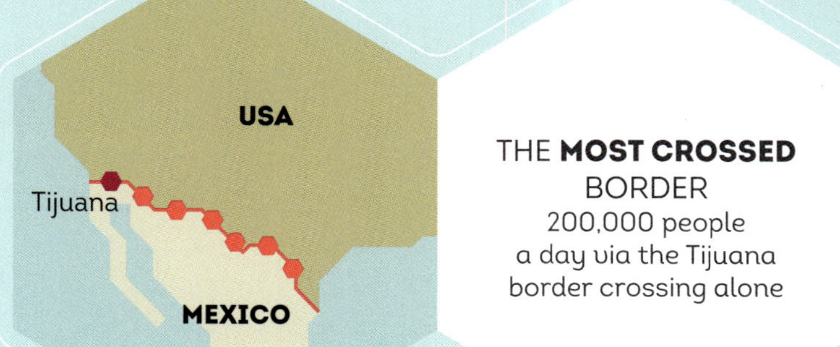

THE **MOST CROSSED**
BORDER
200,000 people
a day via the Tijuana
border crossing alone

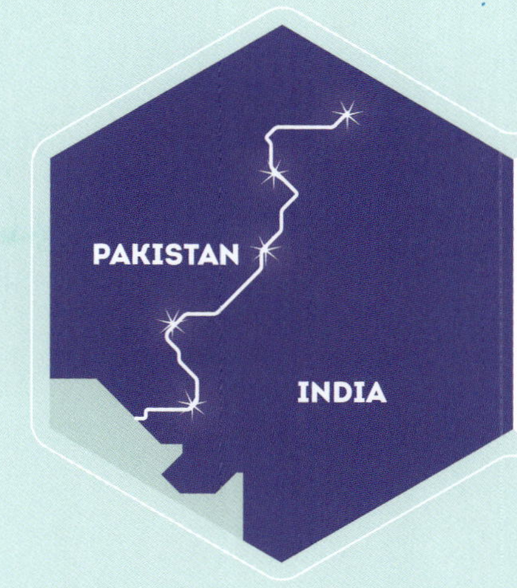

THE ONLY BORDER
VISIBLE FROM SPACE
The line of surveillance posts is visible at night from the International Space Station.

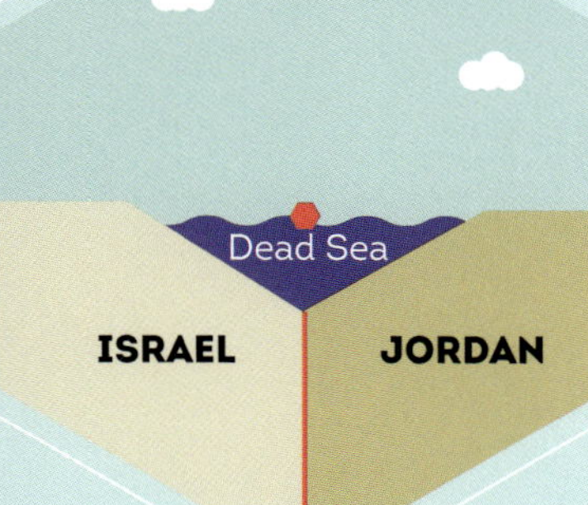

THE **LOWEST** BORDER
-420 m

THE POINT WHERE
THE MOST COUNTRIES' BORDERS MEET
Argentina, Australia, Chile, France, Norway, New Zealand and the UK

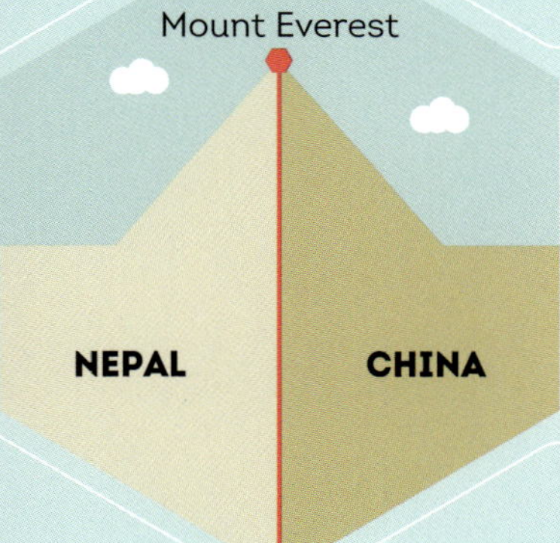

THE **HIGHEST** BORDER
8,848 m

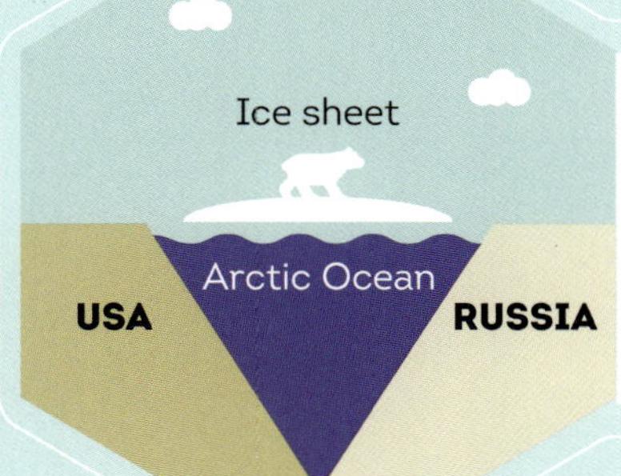

THE DEEPEST TERRITORIAL CLAIM
In 2007, Russia planted its flag 4,261 metres beneath the Arctic Ocean.

Multiple sources.

40 THE INTERNATIONAL DATE LINE

A shift in time

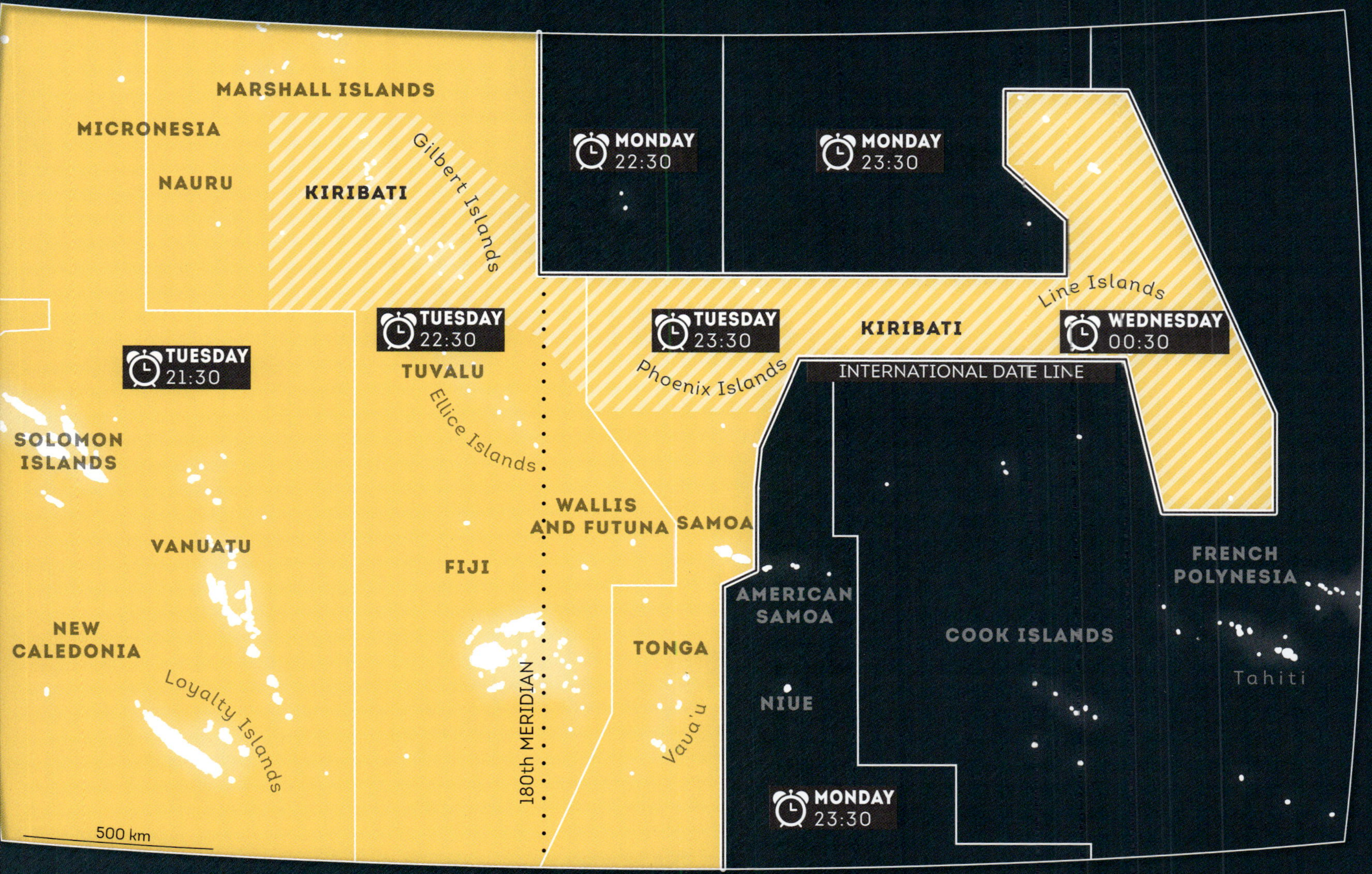

The International Date Line (IDL), which roughly follows the 180° line of longitude – directly opposite the Greenwich Meridian – is often a source of amazement for travellers crossing the Pacific. Although the time on their watch does not change, they find themselves jumping backwards or forwards by a day. The line was arbitrarily fixed at this position because it is very sparsely populated. However, the inhabitants of the Pacific island groups that straddle it can easily travel into the 'past' or the 'future'. One such case is the residents of the islands of Kiribati. Part of their territory lies at GMT+14: this means that when it's 10.30 am on Tuesday in London, and 11.30 pm on Monday on the island nation of Niue, it is already 12.30 am on Wednesday on the 'Line Islands' of Kirabati. The people of those islands are therefore always the first to see in the New Year! The International Date Line can even be moved: in late 2011, the islands of Samoa chose to give up an entire day, in order to switch to the Asian side of the line. But it is in the Bering Strait that the IDL is at its most symbolic: it separates not only two continents, but also two Cold War adversaries and two islands from the same group. Little Diomede (nicknamed 'Yesterday Island') on the American side, and Big Diomede ('Tomorrow Island') on the Russian side, are situated only 3.8 km apart.

Sources: www.fuseau-horaire.com; www.worldatlas.com; 'Comprendre l'étrange découpage des fuseaux horaires', *Le Monde*, 26 December 2014.

AN ICY DESERT PROTECTED FOR SCIENCE ...

More than 100 scientific bases, which include:
- Chinese
- American
- French
- Russian

South Pole Traverse

Marine Protected Area since December 2017

... BUT DEMANDS FOR ACCESS ARE ON THE RISE

Mineral resources
- Oil
- Gas
- Coal
- Metals

Marine resources

Concentration of krill (tiny crustaceans)

Tourism zone

THE ANTARCTIC ICE SHEET IS SHRINKING DUE TO CLIMATE CHANGE

with ice sheet

without ice sheet, land exposed

Borders of territorial claims

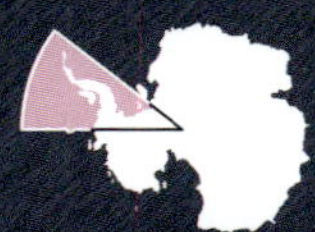

Chile

Argentina

UK

Norway

Unclaimed

New Zealand

France

Australia

SOUTH AMERICA

Pacific Ocean

41

ANTARCTICA

Who controls the South Pole?

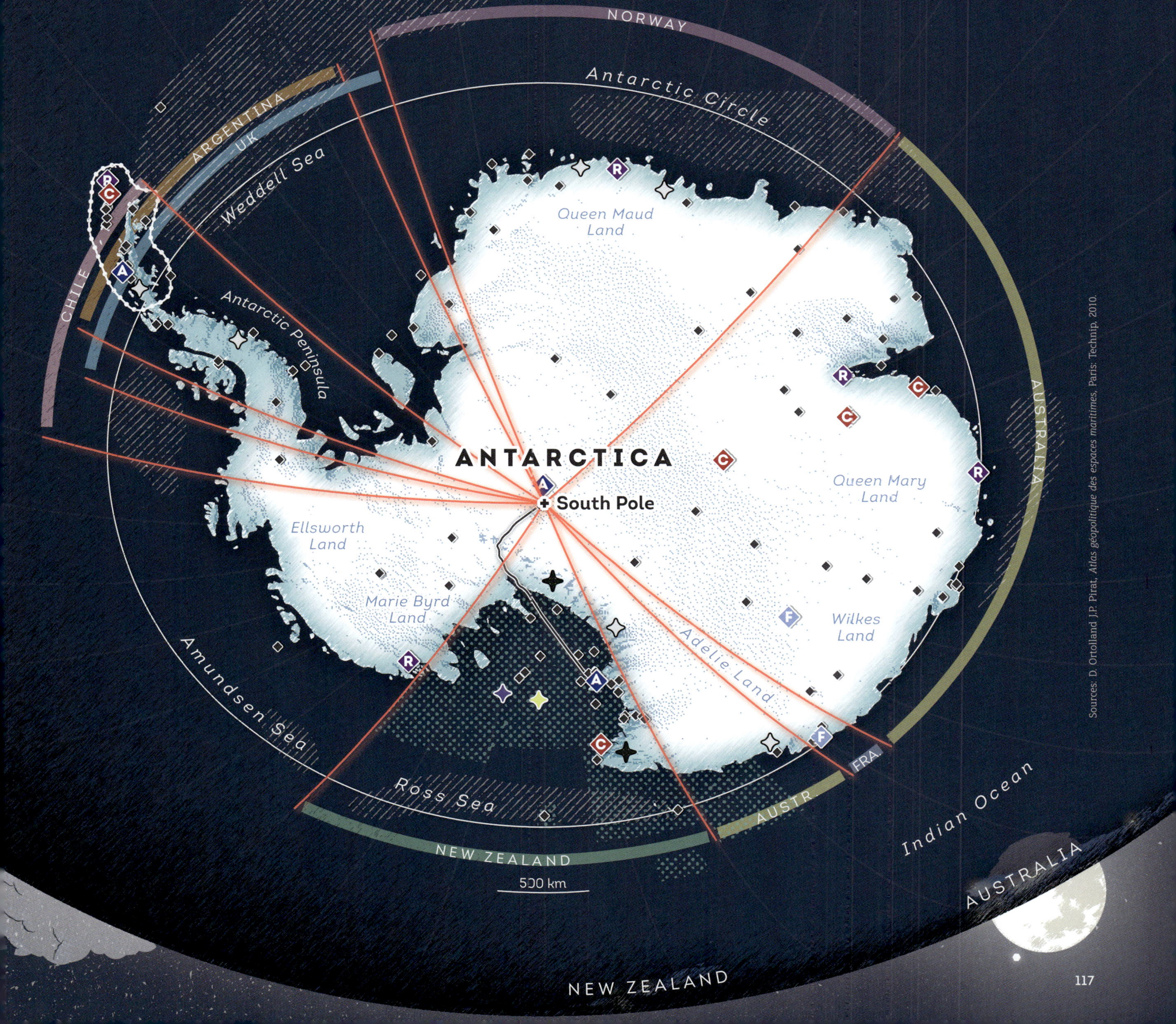

Atlantic Ocean

NORWAY

Antarctic Circle

ARGENTINA

UK

CHILE

Weddell Sea

Queen Maud Land

Antarctic Peninsula

ANTARCTICA

South Pole

Queen Mary Land

Ellsworth Land

Marie Byrd Land

Wilkes Land

Adélie Land

Amundsen Sea

Ross Sea

AUSTRALIA

FRA.

AUSTR.

NEW ZEALAND

500 km

Indian Ocean

AUSTRALIA

NEW ZEALAND

Sources: D. Ortolland J.P. Pirat, *Atlas géopolitique des espaces maritimes*, Paris: Technip, 2010.

42 LANDLOCKED NATIONS

Far from the sea

- **Double landlocked country**, where two borders must be crossed to reach the sea (excepting inland seas)
- Landlocked country, with no access to the coast
- Country with access to the coast of an inland sea
- Country with access to the sea via a river that can be navigated by seagoing ships

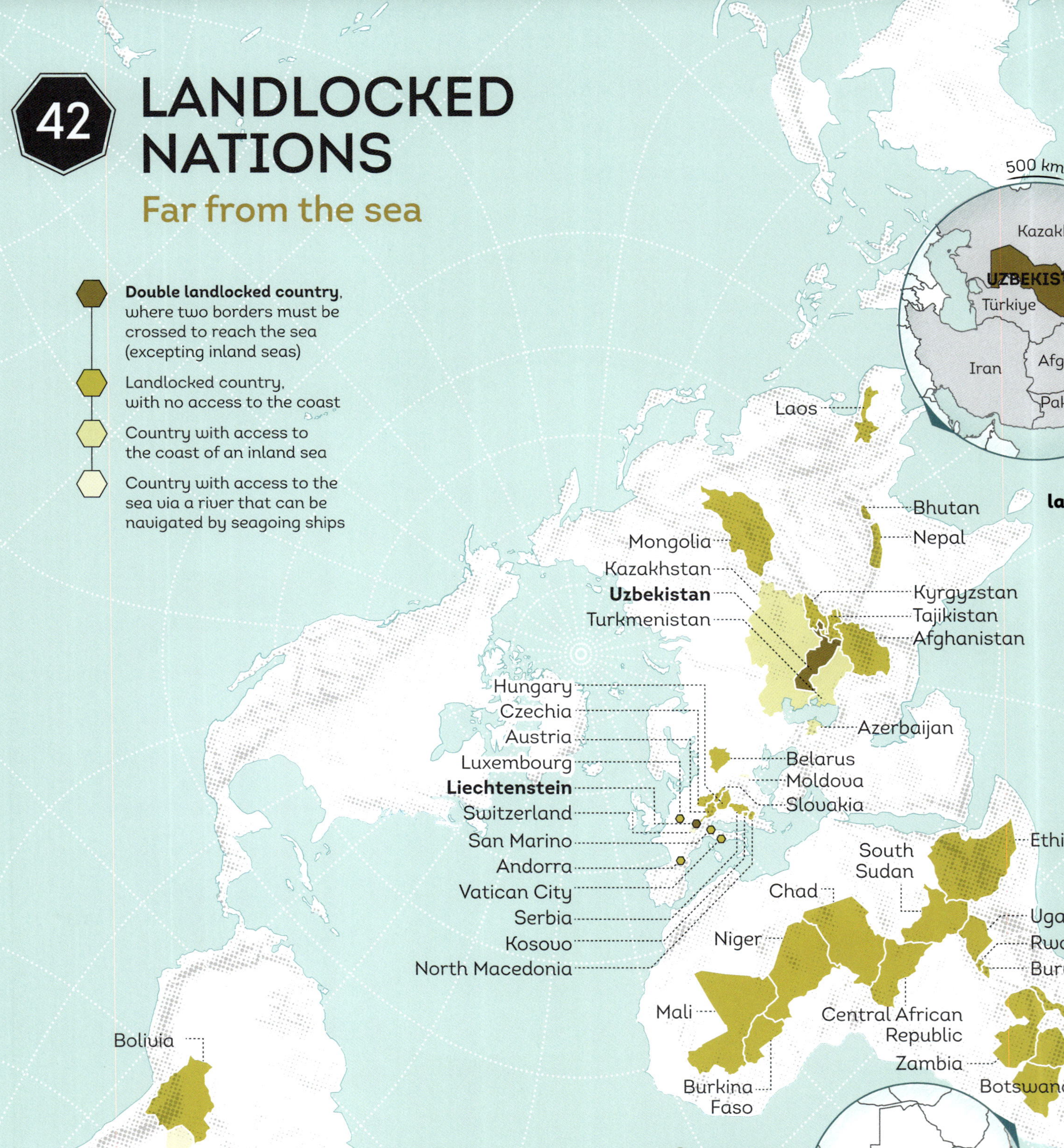

43

WHAT IS A STATE?

The viewpoint of the United Nations

In international law, a state must have a defined territory and a permanent population, and must be self-governing and be internationally recognized. This means that some 'proto-states' possess an uncertain status: the degree of recognition they receive from the UN and the international community may vary.

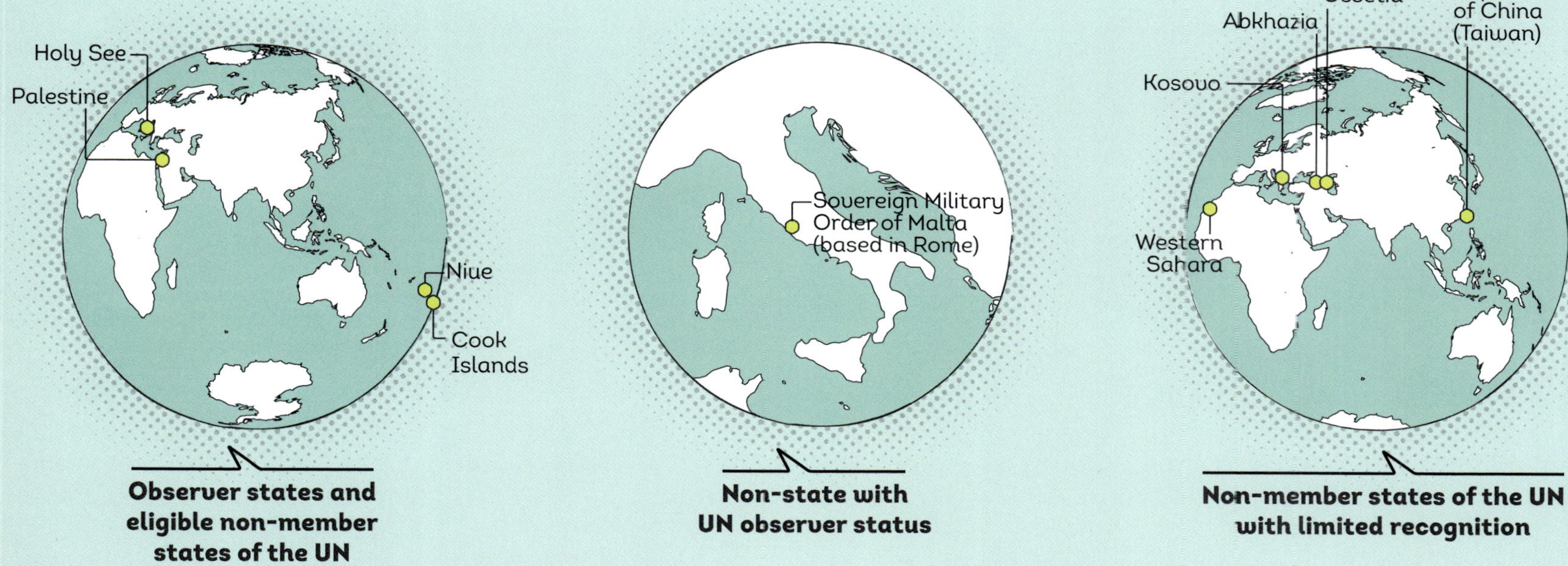

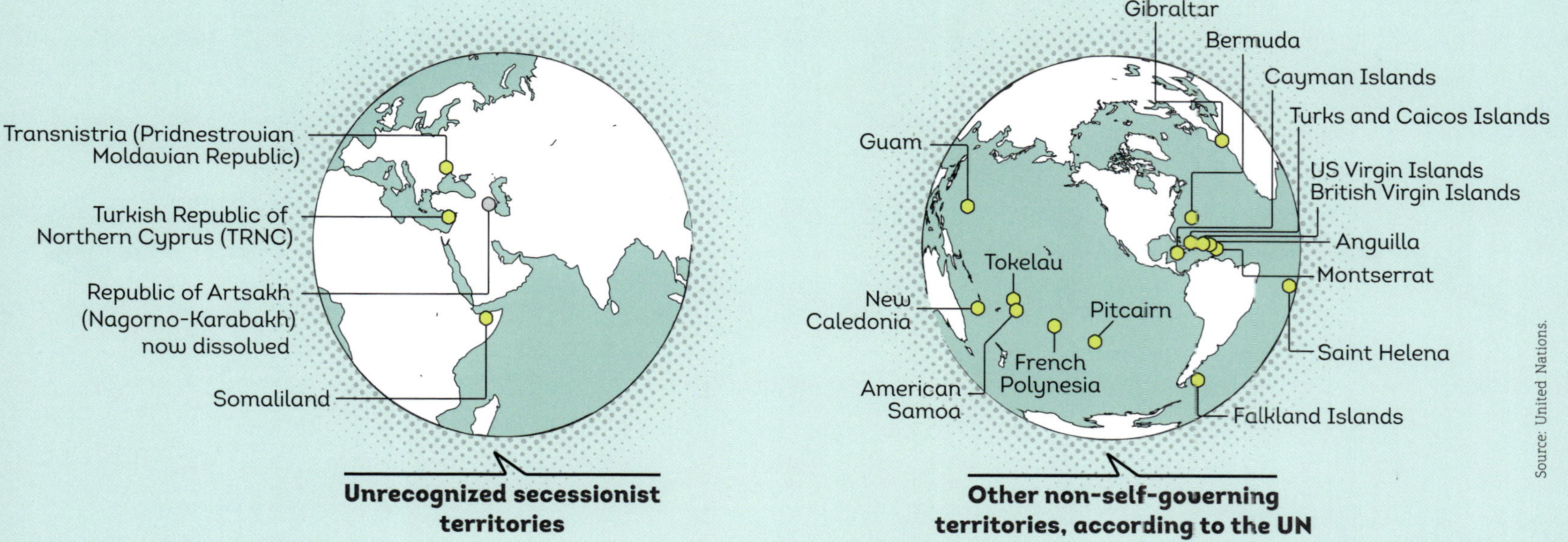

Source: United Nations.

FIVE FORMS OF OCCUPATION

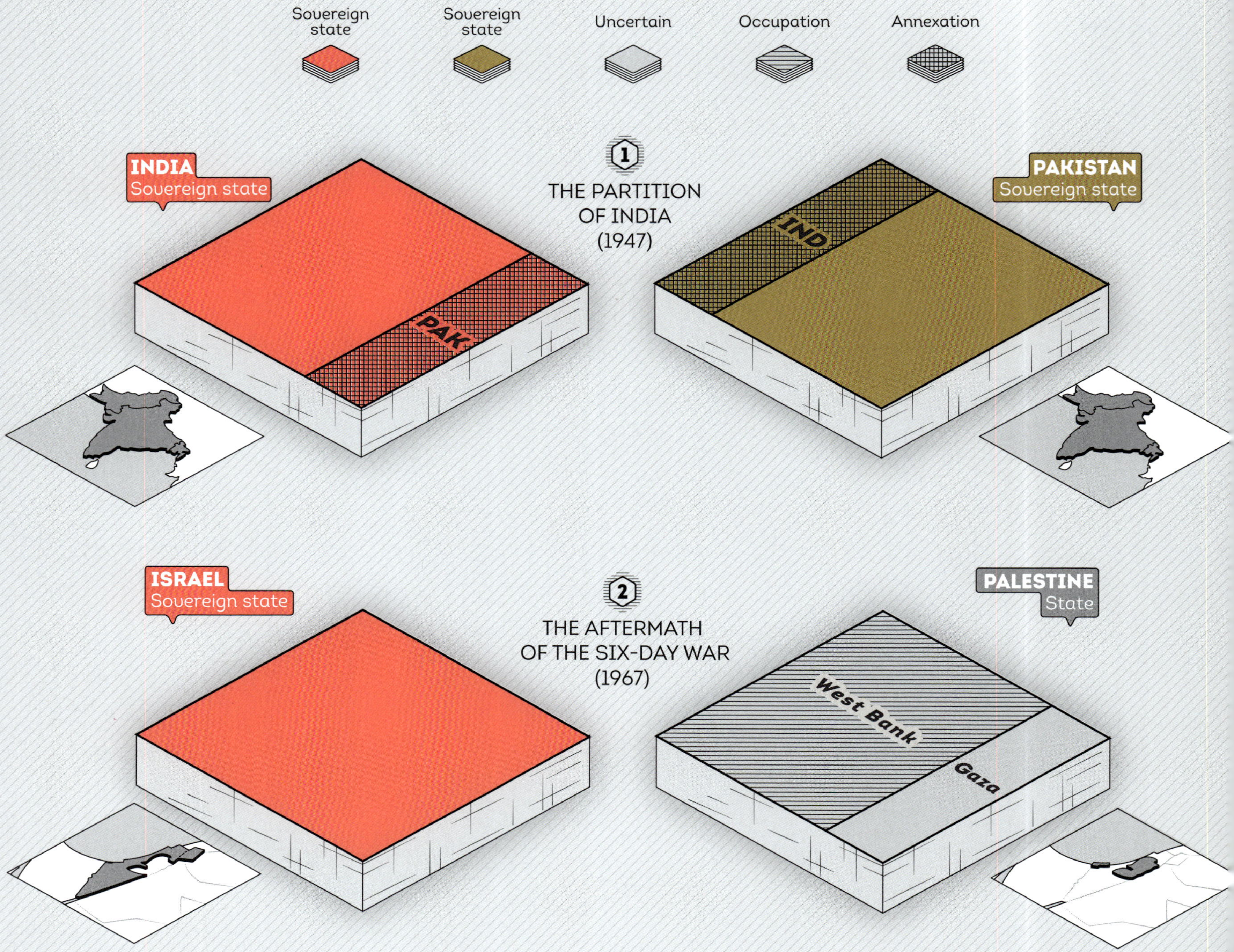

Sovereign state
Sovereign state
Uncertain
Occupation
Annexation

INDIA
Sovereign state

1

THE PARTITION OF INDIA (1947)

PAKISTAN
Sovereign state

PAK

IND

ISRAEL
Sovereign state

2

THE AFTERMATH OF THE SIX-DAY WAR (1967)

PALESTINE
State

West Bank

Gaza

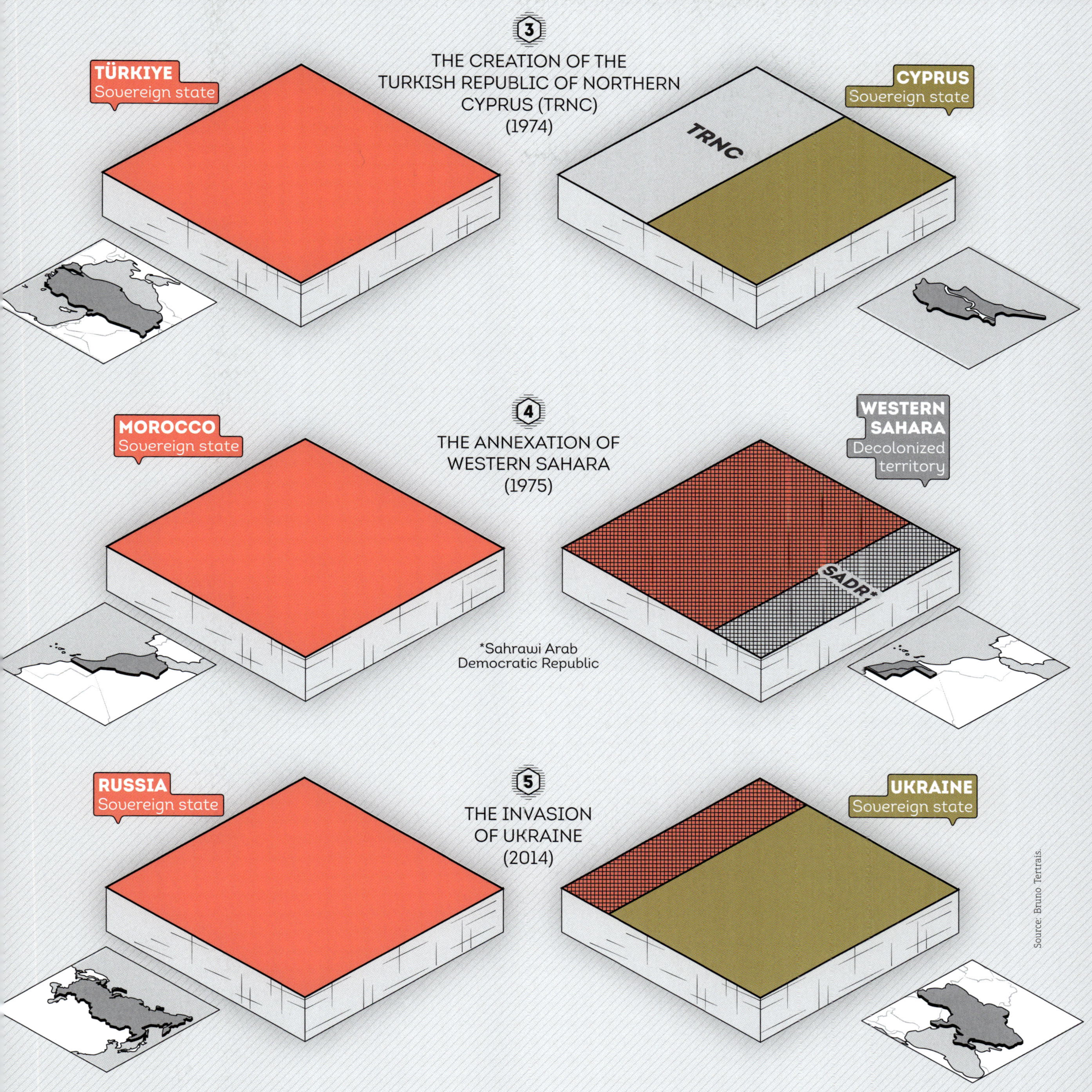
3
THE CREATION OF THE TURKISH REPUBLIC OF NORTHERN CYPRUS (TRNC) (1974)
TÜRKIYE
Sovereign state
CYPRUS
Sovereign state
TRNC
4
THE ANNEXATION OF WESTERN SAHARA (1975)
MOROCCO
Sovereign state
WESTERN SAHARA
Decolonized territory
SADR*
*Sahrawi Arab Democratic Republic
5
THE INVASION OF UKRAINE (2014)
RUSSIA
Sovereign state
UKRAINE
Sovereign state
Source: Bruno Tertrais.

V
DISPUTED BORDERS

Do borders promote peace? Or do they cause conflict? The relationship between borders and conflict is neither simple nor unambiguous. Today, there are around 60 ongoing major territorial conflicts. Some large countries are also seeking to expand their spheres of influence, by force if necessary, far beyond their recognized borders.

V · DISPUTED BORDERS

A border that is recognized by both parties, delimited and – if necessary – demarcated is often born from peace. 'A border also shows that we have made peace,' said Michel Foucher. What's more, it promotes peaceful relations: pairs of countries that have accepted their borders as legitimate and resolved their disputes are very unlikely to go to war. Therefore there is a 'virtuous circle' when it comes to borders.

But the opposite is equally true. Almost all bilateral disputes that have gone through periods of violence have sprung up around borders that are not delimited, borders that were created by decolonization, or secession, or often by a war, for example the dispute between Peru and Bolivia (over sea access). But rather than being a cause of war, borders are more often simply a pretext for it, especially if it is just a question of demarcating the border. It seems unbelievable that the deadly war between Ethiopia and Eritrea was caused by a tiny dispute over the border near the village of Badme. Would Saddam Hussein's Iraq have invaded Iran under the pretext of an erroneous demarcation of the Shatt al-Arab river border, if it had not seen the newly formed Islamic Republic as a major threat? In the words of the geographer Jacques Ancel: 'There are no border problems, only nation problems.' Similarly, Michel Foucher has written: 'There are no problems with borders; there are only problems with relationships between states and people around borders.'

Sharing resources is clearly a sensitive issue when it comes to delimiting borders (Qatar–Bahrain, Sudan–South Sudan). The question becomes especially delicate when two neighbouring countries are exploiting the same fossil fuel deposits: in effect, one country can extract the resources that are found in the subsoil of the other. But when two countries dispute the sovereignty of a maritime region, that is motivated at least as much by nationalism or the strategic importance of the territory as it is by the desire to control its resources.

The border dispute becomes a symbol of nationhood and can easily stir up passions. In India, an unlucky cartographer who makes a mistake when drawing the borders of the country could be fined. In China, passports show the entirety of the 'South China Sea' as under Chinese control. Google Maps decided not to get involved in this dispute: rather than siding with one country or the other, its maps show the border as it is defined by the country where the user is located.

Depending on the circumstances, a border can be a cause or consequence of peace or, conversely, a cause or consequence of war.

A lack of demarcation and surveillance at a border can also be a source of conflict in a different way: by allowing terrorist groups to cross borders. In Operation Barkhane, a counter-insurgency operation, the French army was deployed mainly around the northern borders of countries in the Sahel region.

Therefore, depending on the circumstances, a border can be a cause or consequence of peace or, conversely, a cause or consequence of war.

Disputes and arbitration

One thing is certain: in the modern era, the principle of *uti possidetis* promotes peace. It states that the legitimate borders of a state declaring independence are those of the territory that it represented as an administrative region within another country or empire. First appearing in the 19th century with the first rulings from international courts, and regularly applied since 1945, the principle was officially adopted by the International Court of Justice (ICJ) in 1986. When it is applicable, it constitutes the simplest and most effective way of preventing or resolving a border conflict.

Rather than being a cause of war, borders are more often simply a pretext for it.

Arbitration has become a common way to resolve disputes. Since the end of the Second World War, the ICJ has ruled on 41 cases, two of which set precedents (Cameroon–Nigeria and Thailand–Cambodia). States can also appeal to the Permanent Court of Arbitration in The Hague, which is not a judicial court but an intergovernmental organization, with a wider remit than the ICJ; Slovenia and Croatia have taken their dispute over the Gulf of Piran to the PCA. The United Nations Convention on the Law of the Sea (UNCLOS) also established an International Tribunal for the Law of the Sea, although this is not often used.

45 BORDER DISPUTES

Conflicts with neighbours

Border disputes are often the legacy of history (decolonization, independence, unresolved conflicts) but some are more recent in origin (maritime disputes sparked by the discovery of natural resources). The major ones – which involve true territorial disputes, not simply disputes around demarcation – number around sixty, the oldest being in Latin America (18th century), while the part of the world least impacted is Western Europe. These border disputes represent more than half of major bilateral political disputes. More than a third have, at one time or another, seen violence erupt. Their origins are very varied: the rejection of colonial borders, the separation of communities, errors or imprecisions in treaties, borders that were never delimited after a ceasefire. But many of these disputes remain unresolved. Some are peaceful: on Hans Island, the Danes and the Canadians took it in turns to plant national flags and leave behind a bottle of whisky or brandy. Others are more tense: the delimitation of maritime borders often stokes rivalries, at a time where nations are increasingly seeking to assert their power and exploit natural resources.

* Dispute settled since 2021

NORTH & SOUTH AMERICA

1 | Canada / Denmark (Hans Island)*
2 | Canada / France (Saint Pierre and Miquelon)
3 | Canada / USA (Machias Seal Island)
4 | Guatemala / Belize
5 | Nicaragua / Colombia (continental shelf)*
6 | Nicaragua / Colombia (San Juan River)
7 | Guyana / Venezuela
8 | France / Suriname (islands in the Maroni River)
9 | Chile / Bolivia (waters of the Silala River)*
10 | Argentina / UK (Falkland Islands)

EUROPE

11 | Russia / Estonia
12 | Germany / Austria / Switzerland
13 | Slovakia / Hungary
14 | Russia / Ukraine
15 | Ukraine / Romania (Maican Island)
16 | Armenia / Azerbaijan
17 | Georgia / Russia
18 | Greece / Türkiye
19 | Spain / Portugal (Olivenza)
20 | France / Italy (Mont Blanc)
21 | Italy / Austria (South Tyrol)
22 | Slovenia / Croatia (Gulf of Piran)
23 | Spain / Morocco
24 | Spain / Portugal (Selvagens Islands)
25 | Spain / UK (Gibraltar)

MIDDLE EAST

26 | Israel / Lebanon (maritime border, agreement signed in 2022)*
27 | Israel / Lebanon (land border)
28 | Iran / United Arab Emirates

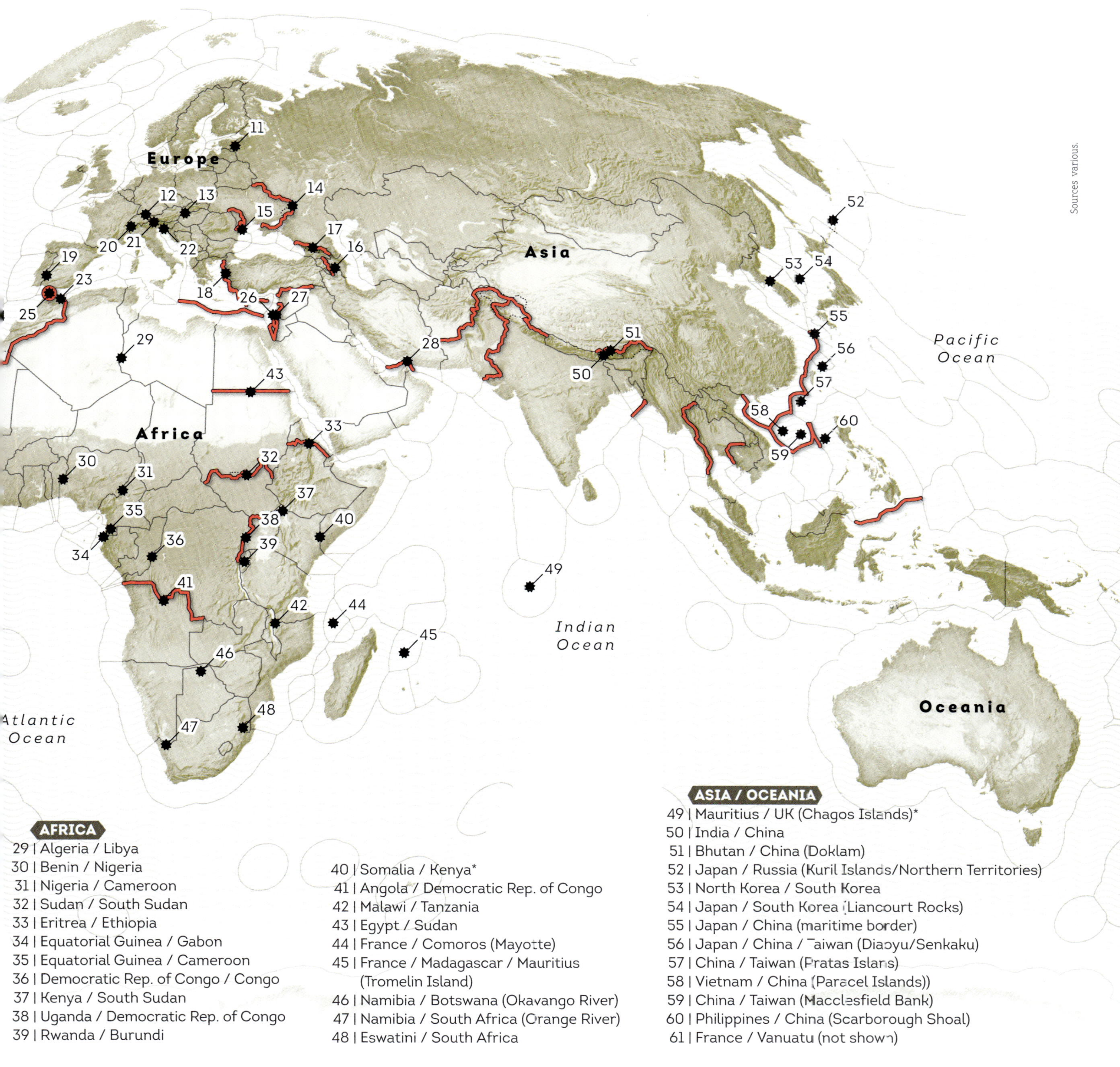

Europe
Asia
Africa
Oceania
Pacific Ocean
Indian Ocean
Atlantic Ocean
Sources various.
AFRICA
29 | Algeria / Libya
30 | Benin / Nigeria
31 | Nigeria / Cameroon
32 | Sudan / South Sudan
33 | Eritrea / Ethiopia
34 | Equatorial Guinea / Gabon
35 | Equatorial Guinea / Cameroon
36 | Democratic Rep. of Congo / Congo
37 | Kenya / South Sudan
38 | Uganda / Democratic Rep. of Congo
39 | Rwanda / Burundi
40 | Somalia / Kenya*
41 | Angola / Democratic Rep. of Congo
42 | Malawi / Tanzania
43 | Egypt / Sudan
44 | France / Comoros (Mayotte)
45 | France / Madagascar / Mauritius (Tromelin Island)
46 | Namibia / Botswana (Okavango River)
47 | Namibia / South Africa (Orange River)
48 | Eswatini / South Africa
ASIA / OCEANIA
49 | Mauritius / UK (Chagos Islands)*
50 | India / China
51 | Bhutan / China (Doklam)
52 | Japan / Russia (Kuril Islands/Northern Territories)
53 | North Korea / South Korea
54 | Japan / South Korea (Liancourt Rocks)
55 | Japan / China (maritime border)
56 | Japan / China / Taiwan (Diaoyu/Senkaku)
57 | China / Taiwan (Pratas Islands)
58 | Vietnam / China (Paracel Islands))
59 | China / Taiwan (Macclesfield Bank)
60 | Philippines / China (Scarborough Shoal)
61 | France / Vanuatu (not shown)

46 THE HORN OF AFRICA

A fragmented future?

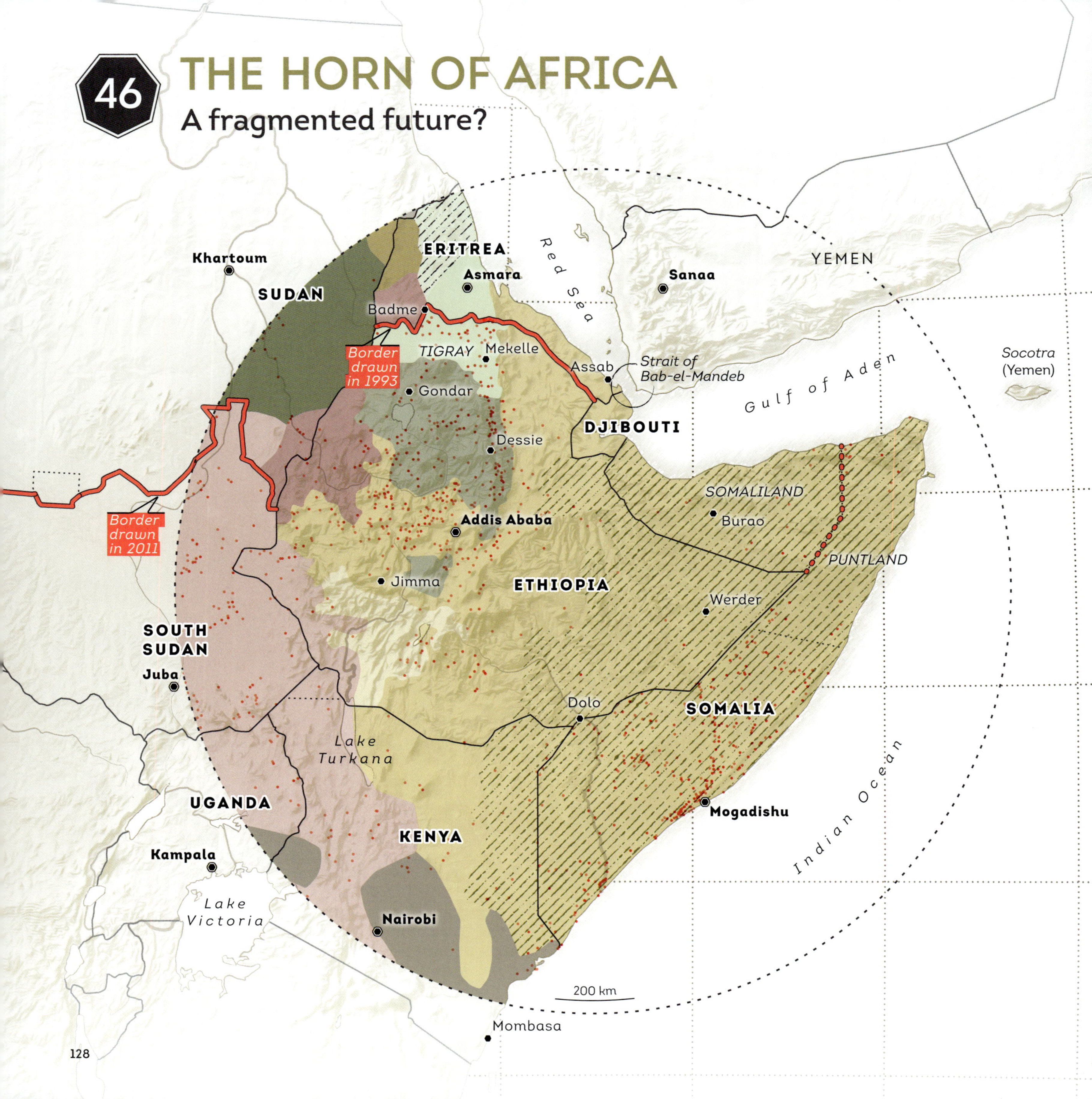

NEW BORDERS AND REGIONAL TENSIONS

New border created after a territorial secession that resulted in the formation of a new state. These two exceptional situations call into question the principle of the intangibility of borders created by decolonization (Cairo Declaration by the Organization of African Unity, July 1964)

Border of self-proclaimed but unrecognized independent state of Somaliland (since 1991)

Violent incidents causing at least five deaths, between 2020 and 2024

A LINGUISTIC PERSPECTIVE

Afro-Asiatic languages

Semitic

- Amharic
- Arabic
- Tigrinya
- Tigre

Cushitic

- Beja
- Other
- Oromo
- Somali

Nilo-Saharan languages

- Eastern Sudanic
- Other

Niger-Congo languages

- Bantu

For many decades, the Horn of Africa, a region of great ethnic diversity, has experienced instability around territory and borders, partly due to the way the land was carved up after decolonization. The former Italian colony of Eritrea, which became part of the Ethiopian Empire in 1952, seceded from Ethiopia in 1993; it fought a deadly war with its neighbour in the late 1990s. This was not resolved until 2018, and trouble flared up again in the Tigray region in 2020. The French Territory of the Afars and the Issas became the independent nation of Djibouti in 1977. The independent nation of Somalia (1960) included the former protectorates of Italian Somaliland and British Somaliland; the latter region then seceded in 1991 and declared itself the Republic of Somaliland. Ethiopia and Somalia historically have a very tense relationship, with disputes over Ogaden in the 1960s and 1970s. The history of Anglo-Egyptian Sudan, which became independent Sudan in 1956, is also littered with bloody civil wars. South Sudan seceded from Sudan in 2011.

Sources: M. Foucher, *Frontières d'Afrique: Pour en finir avec un mythe*, Paris: CNRS Éditions, 2014; J. Sellier, *Atlas des peuples d'Afrique*, Paris: La Découverte, 2011

47 SUDAN AND SOUTH SUDAN

A new border that's already disputed

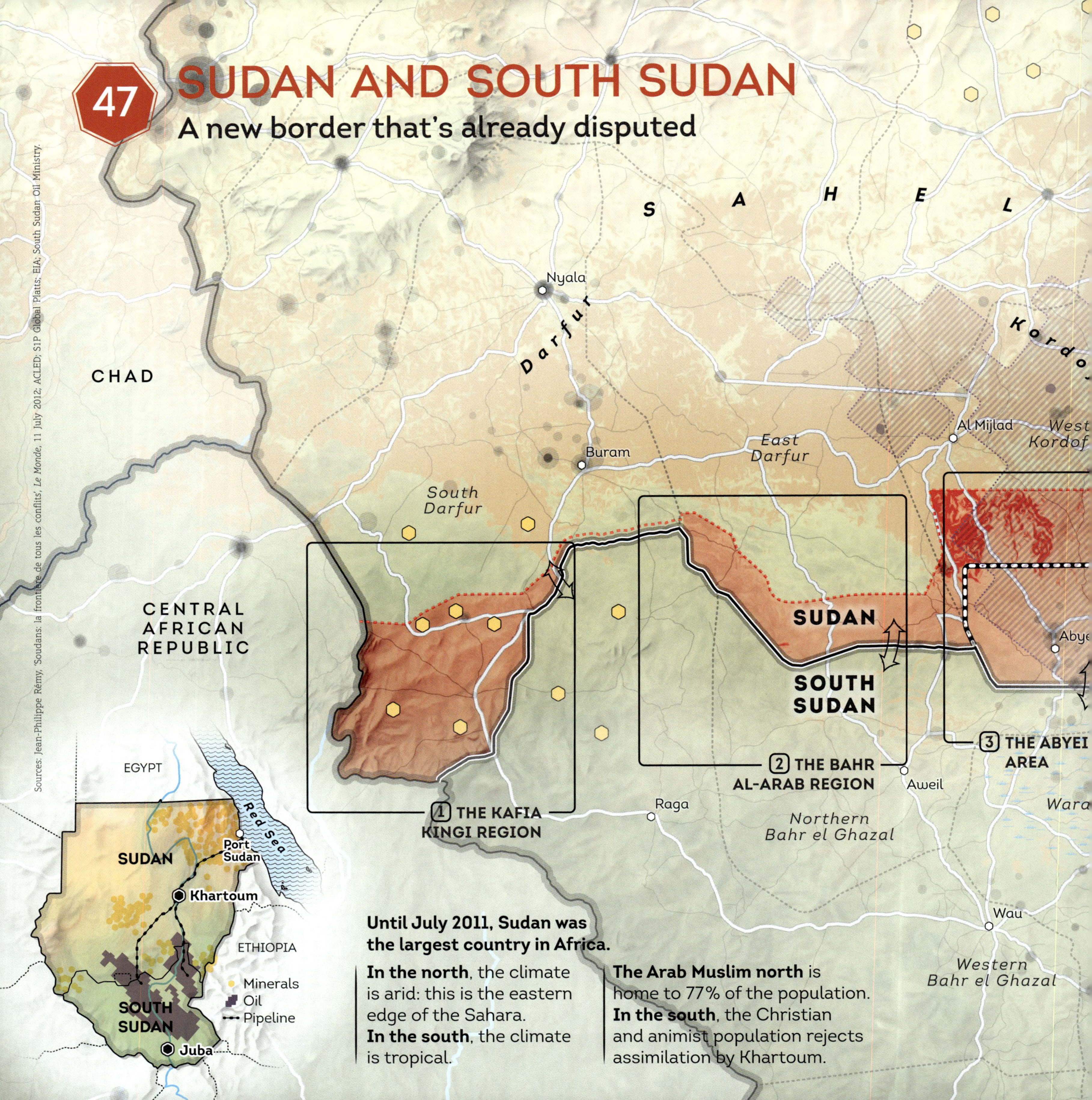

Until July 2011, Sudan was the largest country in Africa.

In the north, the climate is arid: this is the eastern edge of the Sahara. **In the south**, the climate is tropical.

The Arab Muslim north is home to 77% of the population. **In the south**, the Christian and animist population rejects assimilation by Khartoum.

Sources: Jean-Philippe Rémy, 'Soudans: la frontière de tous les conflits', *Le Monde*, 11 July 2012; ACLED; S1P Global Platts; EIA; South Sudan Oil Ministry.

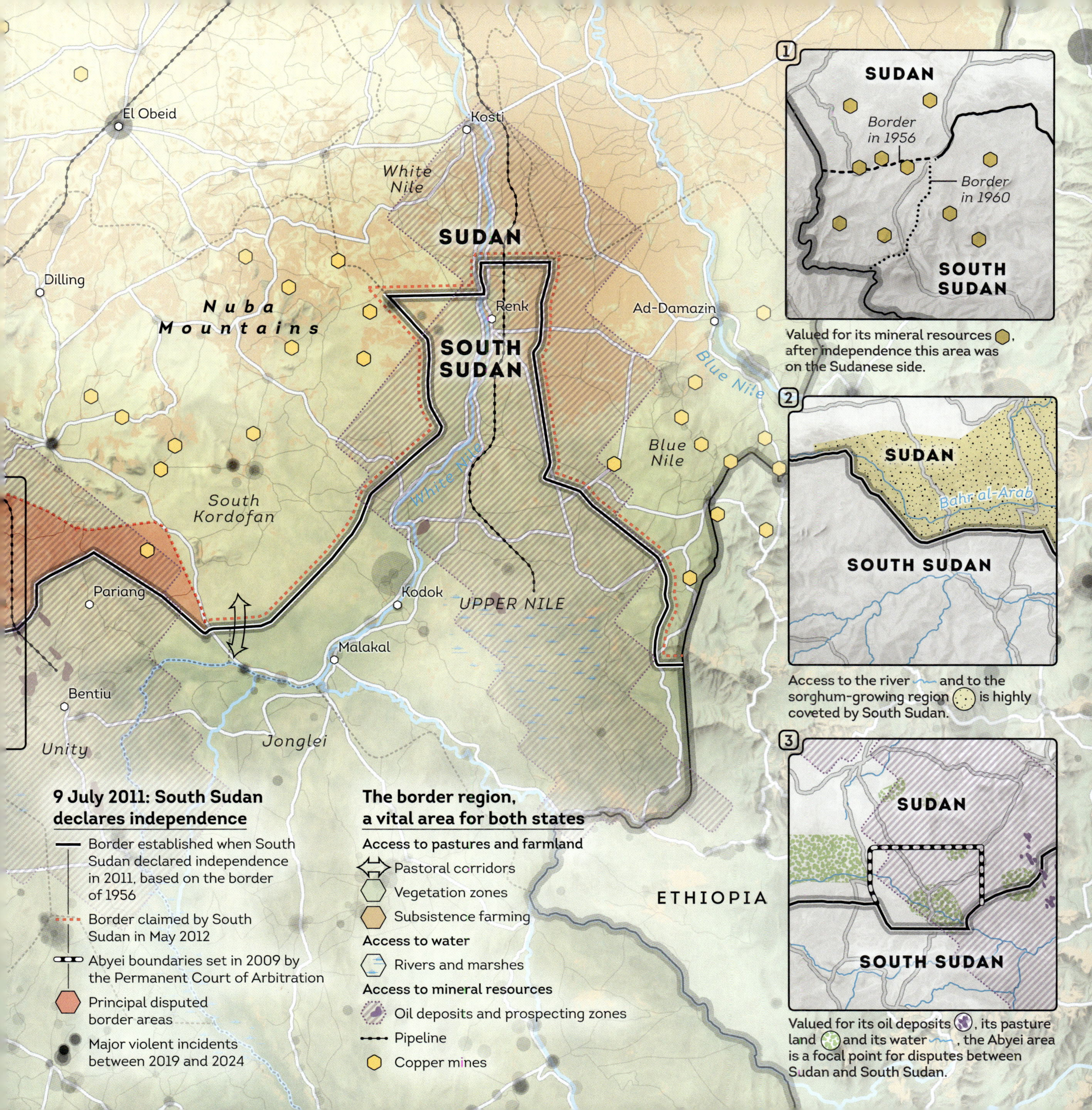
El Obeid
Kosti
White Nile
SUDAN
Dilling
Nuba Mountains
Renk
SOUTH SUDAN
Ad-Damazin
Blue Nile
Blue Nile
White Nile
South Kordofan
Pariang
Kodok
UPPER NILE
Malakal
Bentiu
Unity
Jonglei
ETHIOPIA
9 July 2011: South Sudan declares independence
Border established when South Sudan declared independence in 2011, based on the border of 1956
Border claimed by South Sudan in May 2012
Abyei boundaries set in 2009 by the Permanent Court of Arbitration
Principal disputed border areas
Major violent incidents between 2019 and 2024
The border region, a vital area for both states
Access to pastures and farmland
Pastoral corridors
Vegetation zones
Subsistence farming
Access to water
Rivers and marshes
Access to mineral resources
Oil deposits and prospecting zones
Pipeline
Copper mines
1
SUDAN
Border in 1956
Border in 1960
SOUTH SUDAN
Valued for its mineral resources, after independence this area was on the Sudanese side.
2
SUDAN
Bahr al-Arab
SOUTH SUDAN
Access to the river and to the sorghum-growing region is highly coveted by South Sudan.
3
SUDAN
SOUTH SUDAN
Valued for its oil deposits, its pasture land and its water, the Abyei area is a focal point for disputes between Sudan and South Sudan.

48 THE SAHEL REGION

Borders in the desert

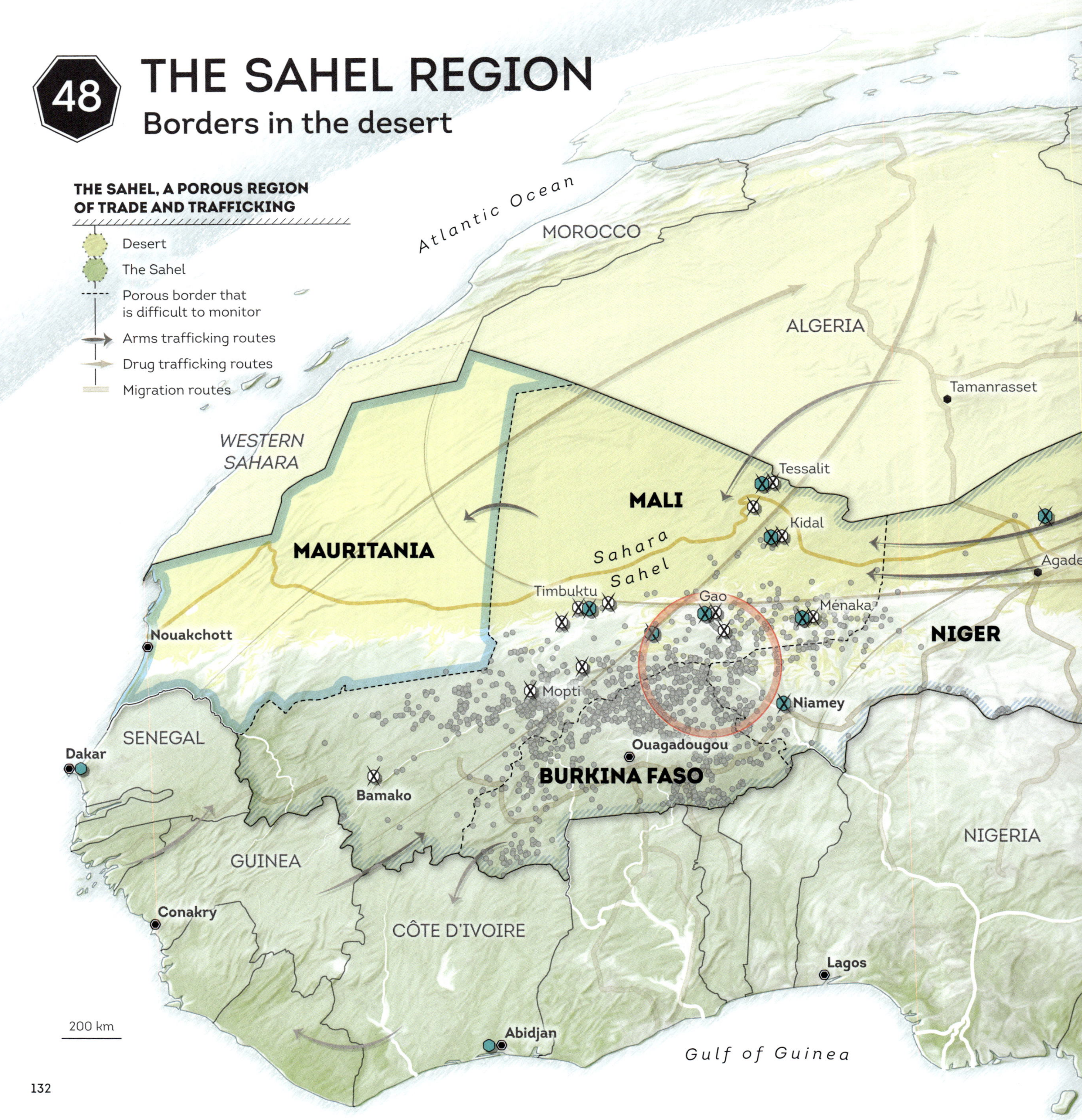

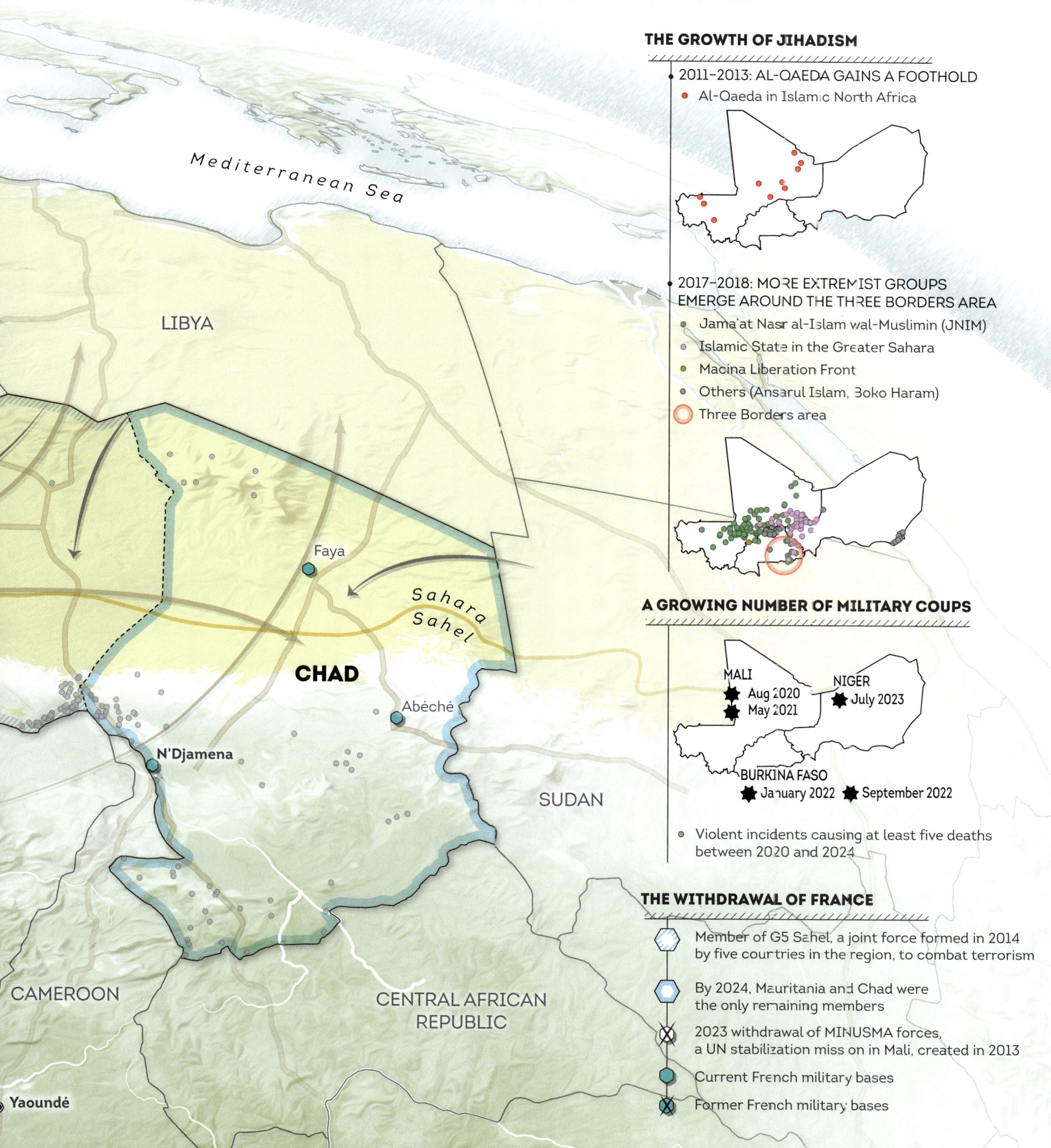

Sources: Armed Conflict Location & Event Data (ACLED); World Atlas of Illicit Flows, Interpol, Rhipto, The Global Initiative Against Transnational Organized Crime, 2019; World Bank; United Nations; Crisis Group; *Le Monde*, 11 October 2020.

THE WEST BANK

A contested border

Israel's West Bank wall, also known as the 'separation barrier', was originally designed to stop terrorists from entering the country. The barrier runs close to the Green Line, sometimes called the '1967 border', although this is incorrect in two ways: it actually dates back to 1949 and it is the ceasefire line, which did not legally settle the two parties' conflicting claims. Only around 10% of the boundary is walled, in areas that are dangerous (gunfire) or densely populated areas where it is impossible to create a separation corridor. Conceived by Yitzhak Rabin's government and built by Ariel Sharon's administration, the wall has broad support within Israel. Its complexity is illustrated by its length: once finished – at more than 700 km – it will be twice as long as the Green Line, which it only follows for 20% of its route. It is true that it dips into the West Bank to encircle two settlement blocs and extends beyond Jerusalem. In total, around 550 km of wall have been built in the West Bank, although between 17% and 9% of the land included within the barrier by the Israeli Supreme Court lies beyond the Green Line. (In 2004, the International Court of Justice ruled that the wall's placement was illegal where it went beyond the Green Line.) This expropriated land is 'temporarily requisitioned', as the term is understood in Ottoman law. One accusation levelled at the wall is that it makes the lives of Palestinians much more complicated; it also gives physical form to a future border without explicitly creating one. This criticism has been made by both the Palestinian Authority and the Israeli far right, which is against the creation of a Palestinian state.

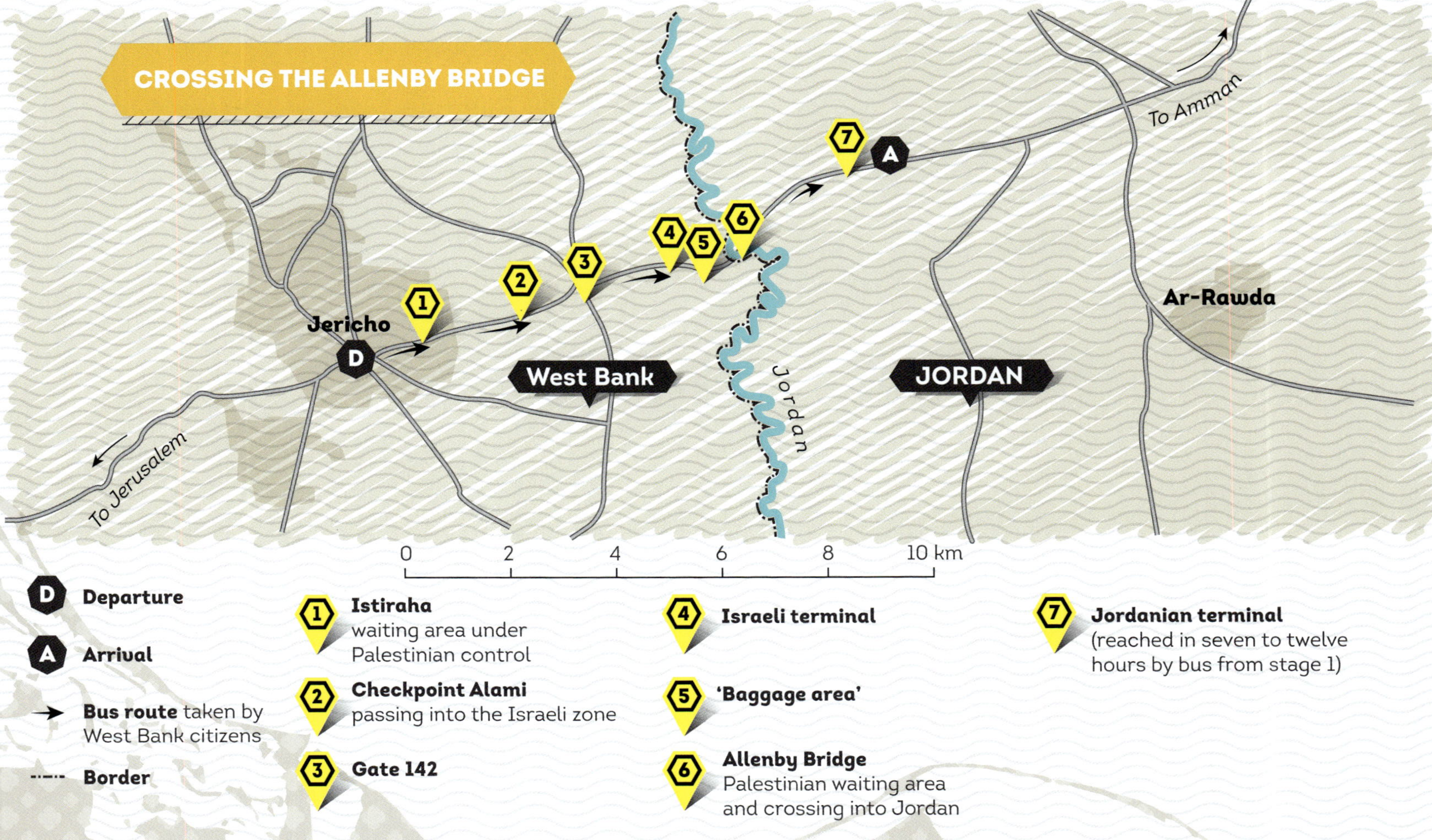

THE ISRAELI WALL
The Green Line
Armistice line dating to 1949
Israeli security wall
constructed or planned since 2002 to defend against terrorist attacks
Seam Zone on the Israeli side
Around 9% of the land (31,000 Palestinians) once the wall is finished
A STRICTLY CONTROLLED EXIT IN THE EAST
Allenby Bridge, the only exit point for Palestinians travelling towards Jordan, is subject to a series border controls run by Israeli, Palestinian and Jordanian authorities (see the detail map opposite).
Irbid
Jordan
Nablus
Tel Aviv
West Bank
Amman
Ramallah
Allenby Bridge
Latrun salient
Jerusalem
JORDAN
ISRAEL
Mediterranean Sea
Dead Sea
Hebron
Gaza
Gaza Strip
Palestinian territories
20 km
EGYPT
Sources: US Government 'Vision for Peace', 2020; OCHA; F. Encel, *Atlas géopolitique d'Israël*, Paris: Autrement, 2018.

PALESTINE

A state whose existence is questioned

The situation born of the ceasefire agreements in 1949 created three territories: Israel; the Gaza Strip, under Egyptian occupation; and the West Bank, under Jordanian occupation and separated from Israeli territory by the Green Line. After the Six-Day War, Israel occupied Gaza and the West Bank. The Oslo Accords (1993) established three areas: Area A, under Palestinian administrative and security control; Area B, under Israeli security control and Palestinian administrative control; Area C, under Israeli administrative and security control. Israel withdrew from Gaza in 2005 and Hamas took power in 2007. The US plan proposed in 2020 envisaged the creation of a Palestinian state including the West Bank and Gaza – connected by a corridor under Israeli control – which would involve a significant redivision of territory, an exchange of territories, the creation of Israeli enclaves (settlements) in the West Bank and the annexation of the Jordan Valley by Israel.

1949–1993

- State of Israel
- Territories occupied by Egypt and Jordan, then by Israel, from 1967 onwards
- Ceasefire line of 1949, respected until 1967. It became the basis of negotiations for a future Palestinian state in the Oslo Accords in 1993
- Golan Heights, Syrian territory occupied by Israel since 1967

In 2024

West Bank

As agreed in the Oslo Accords, this territory is controlled by …

- The Palestinian Authority, Areas A and B (joint control)
- Israel, Area C
- Israeli settlement
- Barrier built by Israel since 2002

Gaza

- Territory controlled by Hamas

Proposal by the Trump administration (2020)

- State of Israel
- State of Palestine
- Israeli enclave

Sources: US Government 'Vision for Peace', 2020; OCHA; F. Encel, *Atlas géopolitique d'Israël*, Paris: Autrement, 2018

7 October 2023: Hamas crosses into Israeli territory

Furthest point reached by Hamas terrorist commandos on Israeli soil

Town or village where Hamas massacred civilians and took hostages

Israeli barrier breached at multiple points by Hamas fighters after they neutralized the alarm system

Egyptian barrier

Israel retaliates heavily in the Gaza Strip

Displacement of the population living in the north of the Gaza Strip on the advice of the Israeli army

Massive bombardment by the Tzahal (Israeli defence forces), followed by Israeli troops entering the Gaza Strip

Closed crossing point

Maritime exclusion zone

Israeli buffer zone

A WAR ON MULTIPLE LEVELS

1 **In the air,** the Israeli air force supports…

2 … troops deployed **on the ground.**

3 **Below ground,** Hamas retains military autonomy.

Mediterranean Sea

Ashkelon

Erez

Gaza Strip

Gaza

Nahal Oz

Deir al-Balah

Re'im

ISRAEL

Magen

Khan Yunis

Rafah

Rafah

Philadelphi Corridor

Kerem Shalom

EGYPT

51 GAZA

The attacks of 7 October 2023

As set out in the Oslo Accords (1993), Israel continued to control access to Gaza's airspace and maritime area. A security barrier, nicknamed the 'Iron Wall', was built along the Israeli border, which has had a 1-km-wide 'buffer zone' since 2001, reinforced after the Israeli retreat in 2005. In the south, Egypt took possession of the Philadelphi Corridor to create a buffer zone. On 7 October 2023, there were only three land crossing points where people could enter or leave Gaza, two of which were controlled by Israel (Erez in the north for people and goods, Kerem Shalom in the south for goods from Egypt) and one controlled by Egypt (Rafah, for people only). Israel disputed the use of the terms 'blockade' (the movement of goods and people remained possible under certain conditions) and 'occupation' (Israeli forces no longer had a permanent presence in the territory). A number of tunnels, used to smuggle goods and by terrorists entering Israel, have also been destroyed along the two borders over the last twenty years. On 7 October 2023, in addition to launching at least 2,500 rockets, more than 2,000 fighters from Hamas and other terrorist groups crossed over into Israeli territory by land (on motorbikes and in pick-up trucks), opening up around thirty gaps in the security barrier, as well as by air (using motorized paragliders), reaching as far as Ofakim, 30 km from the border.

Sources: ISW, Critical Threats; WarMapper; OCHA; 'Guerre Israël-Hamas: six mois après l'attaque du 7 octobre 2023, le bilan de l'offensive israélienne à Gaza', *Le Monde*, 5 April 2024.

THE MAIN STAGES OF THE ISRAELI MILITARY STRATEGY IN GAZA AFTER THE 7 OCTOBER ATTACKS

7 October 2023
Hamas attack on Israeli soil

7 to 27 October 2023
Security and bombing raids

24 to 30 November 2023 **Trêve humanitaire**

from 27 October 2023
Land offensive in the north

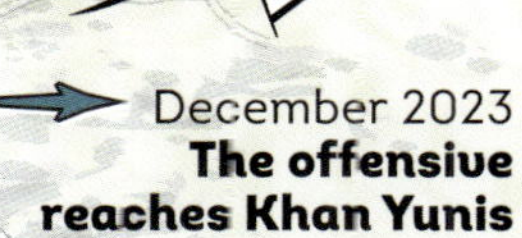

December 2023
The offensive reaches Khan Yunis

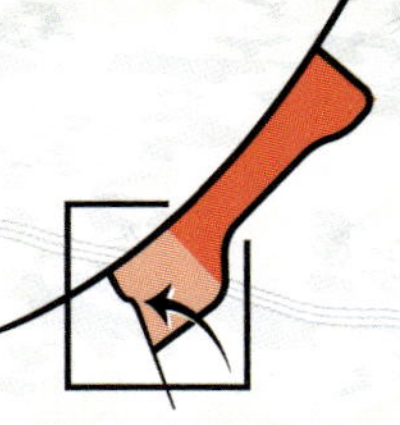

May 2024
Rafah offensive

Ramallah
Migron
Kokhav Ya'akov
Airport
Giv'at Ze'ev
Atarot
WEST BANK
Geva Binyamin (Adam)
Giv'on HaHadasha
Har Shmuel
Neve Ya'akov
Almon
Pisgat Ze'ev
Ramot Allon
Ramat Shlomo
French Hill
Ma'alot Dafna
El Area
Mount Scopus
Mount of Olives
West Jerusalem
Old City
Mount Zion
Ma'ale Adumim
East Jerusalem
ISRAEL
East Talpiot
Givat HaMatos
Gilo
Har Homa
WEST BANK
Rachel's Tomb
Bethlehem
Church of the Nativity
2 km

THE CLOSURE OF JERUSALEM

From security to occupation?

From a geographical perspective, Jerusalem lies in the east of Israel's territory. From a political perspective, the east of the city – conquered during the Six-Day War – was annexed by Israel in 1982, and the city limits therefore stretch beyond the Green Line. From an administrative perspective ('Greater Jerusalem'), the city's surface area has increased tenfold since 1967. The fate of the sector known as 'E1', in the east of the city, is a key issue: here, the security barrier moves a number of kilometres away from the Green Line to surround Israeli settlements including Ma'ale Adumim. Pursuing this policy would make it difficult to establish East Jerusalem as the capital of a Palestinian state. The Arabs who remain there today have the status of 'permanent residents'.

DIVISIONS BETWEEN TWO POPULATIONS

The Green Line (1949–1967): basis of the Oslo negotiations for a future Palestinian state

City limits drawn up by Israel after the Six-Day War in 1967

Security wall, completed or under construction
The Israeli view: This wall protects them from terrorist incursions.
The Palestinian view: This wall cuts off large portions of the West Bank and restricts the free movement of Palestinians.

Sources: United Nations Office for the Coordination of Humanitarian Affairs; Foundation for Middle East Peace; Open Street Map.

THE OLD CITY OF JERUSALEM

Holy ground for multiple faiths

The situation in the old city of Jerusalem echoes the complexity and tensions in the region as a whole. Informally divided into four cultural quarters (Armenian, Christian, Jewish, Muslim) for many years, the old city now has a larger Jewish presence, while al-Aqsa, the 'Esplanade of Mosques' – which is bordered to the west by the Wailing Wall and under the custodianship of the Hashemite rulers of Jordan – is the focal point of tensions. Peace plans proposed by the United States in the 1990s envisaged complex solutions for the place known by Jews as Temple Mount and by Muslims as the Noble Sanctuary. The 1947 proposal put forward by the United Nations, which suggested granting Jerusalem the status of international territory (*corpus separatum*), is unlikely to ever be revived.

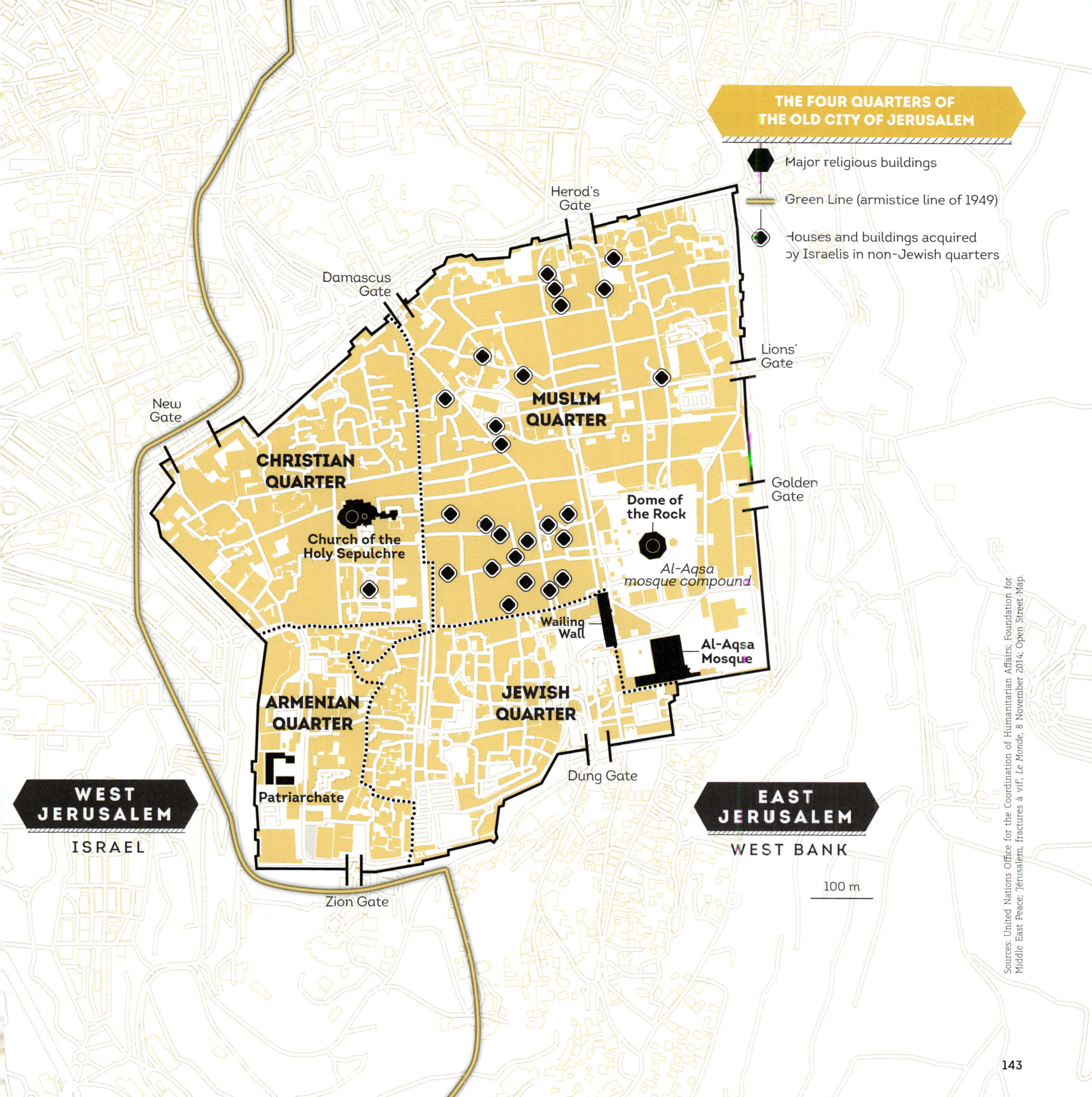

Sources: United Nations Office for the Coordination of Humanitarian Affairs; Foundation for Middle East Peace; 'Jérusalem, fractures à vif', *Le Monde*, 8 November 2014; Open Street Map.

54 THE GOLAN HEIGHTS

A complex border situation

In the Middle East, borders are 'neither certain nor recognized' (Michel Foucher), despite the UN calling for agreements for many decades. To the north of Israel, on its borders with Lebanon and Syria, is one of the most complicated and contentious border regions on the planet. The 'high-tech barrier' (known up until 2000 as the 'Good Fence') built by Israel to give physical form to its border with Lebanon is mostly situated a few dozen metres behind the Blue Line, the border defined by the UN in 2000, which more or less follows the ceasefire line of 1949 (which, in turn, follows the 1923 line defining the border of Mandatory Palestine). However, as well as the Syrian area of Golan Heights, Israel has annexed Shebaa Farms, a small territory to the south of the Blue Line, whose status has never been clarified and which is still claimed by Lebanon, but considered Syrian by the UN. UN troops have patrolled to the north of the border, in Lebanese territory, since 1978, and to the east of Golan Heights, on a strip of demilitarized land, since 1974.

LEBANON
SYRIA
GOLAN
West Bank
JORDAN
Gaza Strip
ISRAEL

A STRATEGICALLY IMPORTANT BORDER TERRITORY

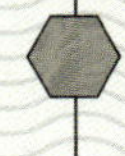
A plateau of strategic significance, at an altitude of over 1,000 m, 60 km from Damascus

Access to the Sea of Galilee: a major source of fresh water for Israel

Main tributaries of the River Jordan and the Sea of Galilee

CONTROLLED BY VARIOUS COUNTRIES DURING THE 20TH CENTURY

Border drawn in 1923 by Great Britain and France during the time of their mandates

Demilitarized zone created in 1949 after the war between Syria and Israel, 1948–1949

Golan Heights taken over by Israel after the Six-Day War in 1967

Security zone occupied by Israel between 1978 and May 2000 (South Lebanon)

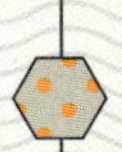
Blue Line, created in 2000 by the UN to mark the Israeli withdrawal

Hezbollah activity

THE REGION IS CURRENTLY OCCUPIED BY ISRAEL

Demilitarized zone under UN control since 1973

UN Interim Force in Lebanon (UNIFIL)

De facto occupation of the Golan Heights by Israel in 1981, recognized by the US in March 2019

Israeli settlements

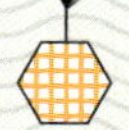
Shebaa Farms:
Syria sees this area as Lebanese territory; Israel sees it as an extension of Golan

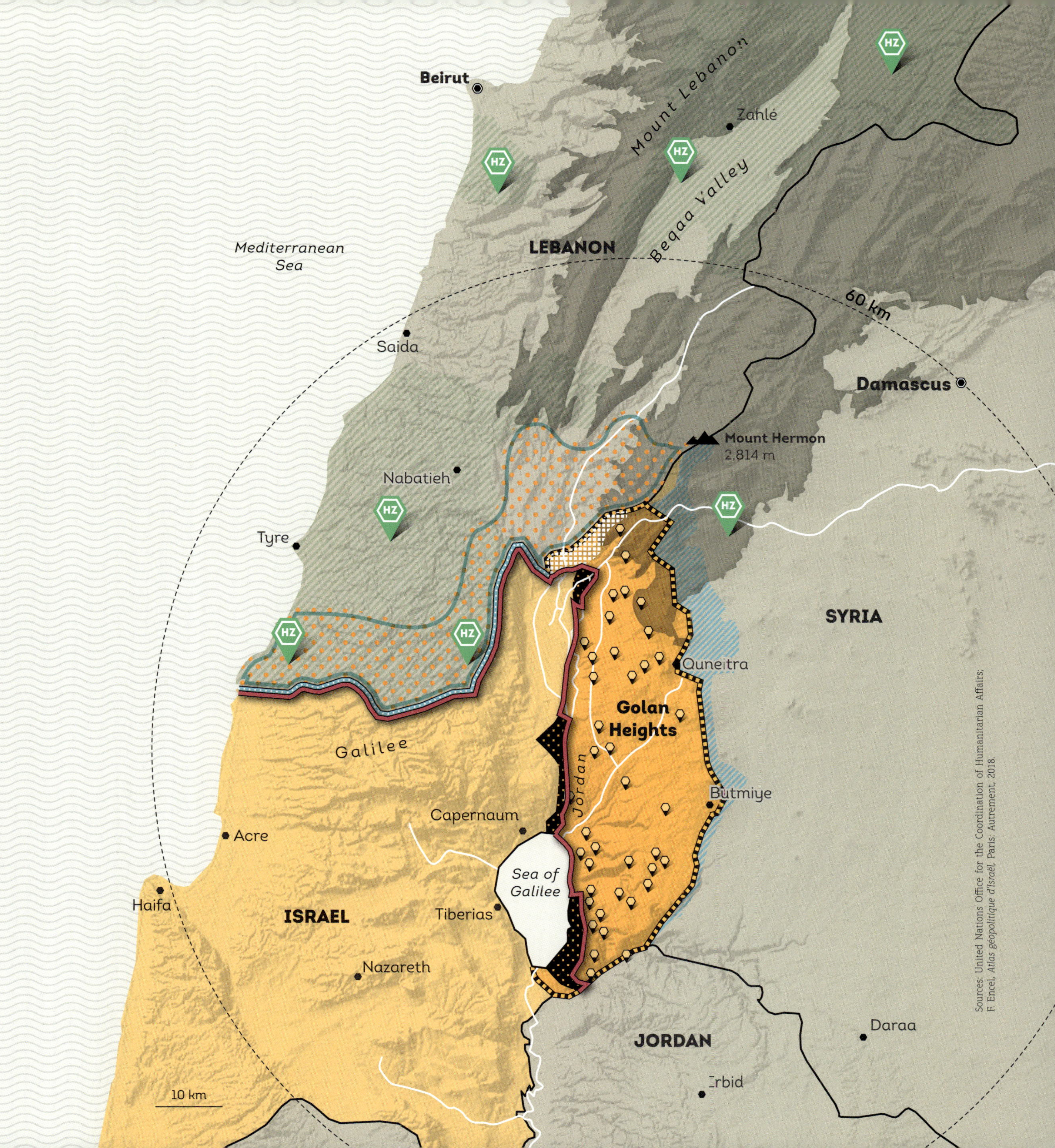

Sources: United Nations Office for the Coordination of Humanitarian Affairs; F. Encel, *Atlas géopolitique d'Israël*, Paris: Autrement, 2018.

55

BORDERS OF THE FORMER YUGOSLAVIA

The scars of division

A WIDELY UNRECOGNIZED STATE

Borders of Kosovo: the country declared independence in 2008. Of the 193 UN member states, 89 still refuse to recognize it, including Serbia, Russia and five countries in the European Union (Greece, Spain, Cyprus, Romania, Slovakia).

ETHNIC GROUPS AND ENCLAVES

- Albanian, majority in Kosovo
- Serb, majority in Serbia
- Other minorities (Roma, Bosnian, etc.)
- Serb enclaves within Kosovo

POTENTIAL LAND SWAPS

- Territorial exchange suggested by the presidents of Kosovo and Serbia in 2018
- From Serbia to Kosovo
- From Kosovo to Serbia

BEFORE 1991

Principal ethnic groups of the Federal Republic of Yugoslavia

- Slovenes
- Hungarians
- Croats
- Muslims
- Serbs
- Montenegrins
- Albanians
- Bulgarians
- Macedonians
- Current borders

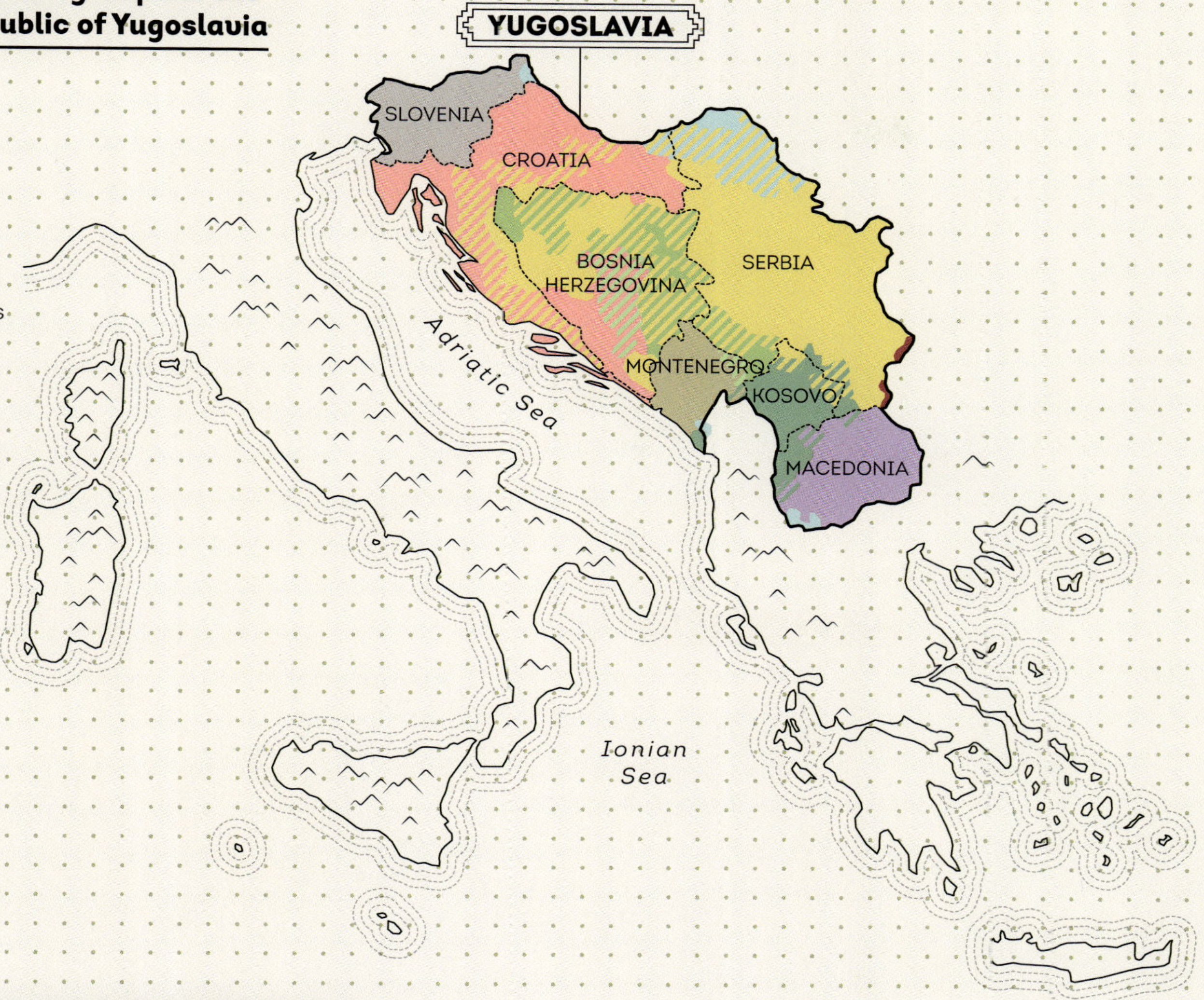

YUGOSLAVIA

SLOVENIA
CROATIA
BOSNIA HERZEGOVINA
SERBIA
MONTENEGRO
KOSOVO
MACEDONIA
Adriatic Sea
Ionian Sea

200 km

Sources: *Le Monde*, 23 September 2018; C. Grataloup, *A History of the World in 500 Maps*, London: Thames & Hudson, 2023.

THE BREAK-UP OF YUGOSLAVIA

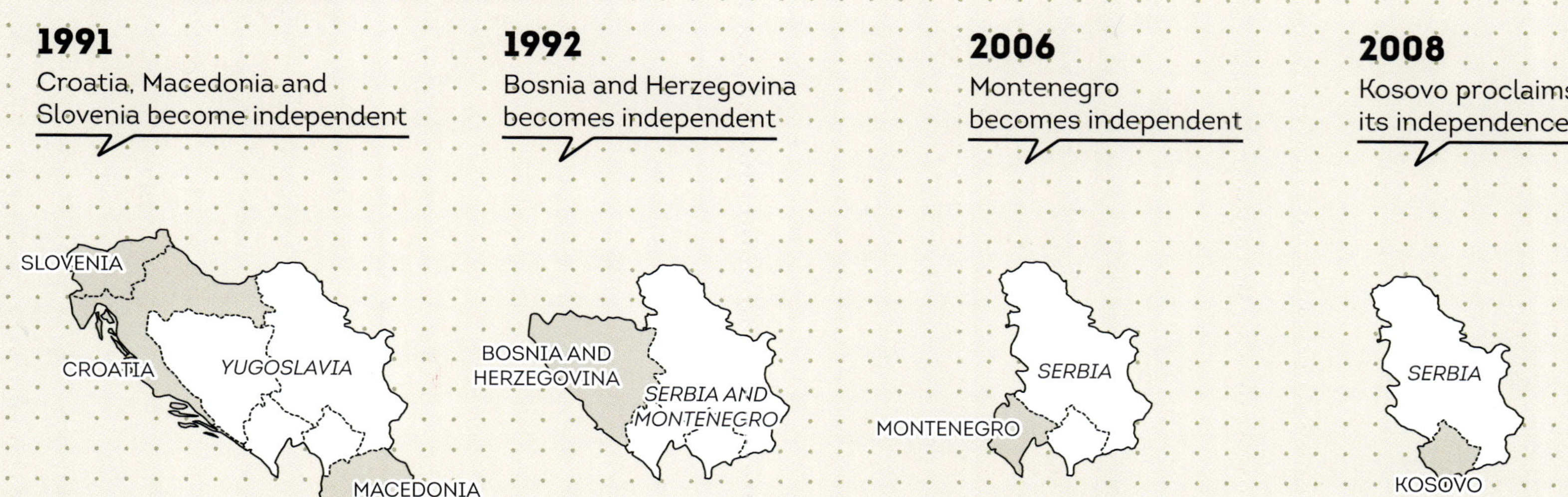

56
GUYANA
Disputes with Venezuela
Dominica
Martinique (Fr.)
Saint Lucia
Barbados
Lesser Antilles
Saint Vincent and the Grenadines
Grenada
Aruba
Netherlands Antilles
Caribbean Sea
ATLANTIC OCEAN
Maritime territory claimed by Venezuela
Trinidad and Tobago
Caracas
Maturín
Roraima Block
Kaieteur Block
Stabroek Block
Kanuku Block
Orinoco Delta
Punta Barima
VENEZUELA
Cordillera de Mérida
Ciudad Bolívar
Ciudad Guayana
Tumeremo
Office issuing Venezuelan national identity cards to people from Essequibo
Georgetown
Ankoko Island
La Camorra
Construction or expansion of Venezuelan military bases
Linden
Essequibo
COLOMBIA
Creation of a landing strip for the Venezuelan military, intended to serve as a 'logistical support point for the integral development of Essequibo'
GUYANA
Guiana Highlands
Boa Vista
BRAZIL
Amazon Rainforest
USA
SOUTH AMERICA
BRAZIL
Equator
Sources: Guyana Ministry of Foreign Affairs; ONSA Venezuela; 'Guyana: l'or noir de l'Essequibo attise les tensions', Le Monde, 24 & 25 March 2024.

POTENTIAL FOSSIL FUEL DEPOSITS IN ESSEQUIBO HAVE REAWAKENED A TERRITORIAL DISPUTE BETWEEN VENEZUELA AND GUYANA

Essequibo is a region measuring 159,500 km² with 125,000 inhabitants. Now part of Guyana, it is **claimed by Venezuela.**

As well as **tropical forests,** the region contains **mineral reserves** (gold, diamonds, copper, uranium, bauxite, iron).

In the early 2000s, Guyana granted foreign multinationals (US, British, Chinese, French) rights to the oil deposits along its coasts.

In 2015, the US company ExxonMobil became the first to discover oil in the Stabroek Block. This discovery attracted the attention of foreign companies and neighbouring Venezuela (maritime claims).

In August 2023, Guyana issued a call for tenders to exploit a number of oil fields. This action was denounced by Venezuela.

On 3 December 2023, Venezuela held a consultation referendum: 95% of Venezuelans supported their nation's claim to Essequibo. The US and British military carried out military exercises in the area as a show of support for Guyana. Meanwhile, Venezuela announced the deployment of soldiers along its border with Guyana.

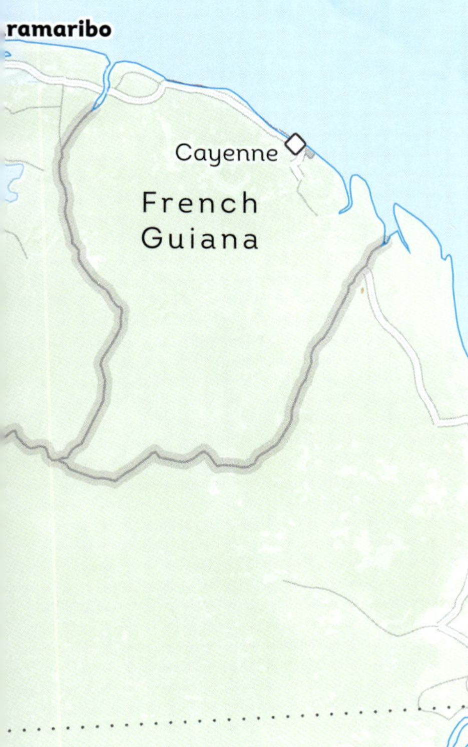

THIS BORDER DISPUTE IS A LEGACY OF THE COLONIAL PAST

Land borders of Guyana

Claimed maritime borders

Current land borders

Claimed maritime borders

Claimed territory

Guyana's perspective

The region of Essequibo was annexed to Guyana in around 1840, when it was still the colony of British Guiana (until 1966). The territorial dispute with Venezuela, supported by the US, was resolved in 1899 by the Court of Arbitration in Paris, which ruled in favour of the British Empire. In 2020, Guyana turned to the International Court of Justice to resolve the dispute. However, objections from Venezuela delayed the final decision, which is still pending.

Venezuela's perspective

Since it was created by the Spanish crown in 1777, Venezuela has considered the Essequibo river to be its 'natural border'. The region was part of its territory when Venezuela became independent in 1811, until the British seized it around 1840. Venezuela rejected the 1899 decision by the Court of Arbitration in Paris. In 1966, the United Kingdom and Guyana acknowledged Venezuela's claim and made a commitment to peacefully resolve the territorial dispute.

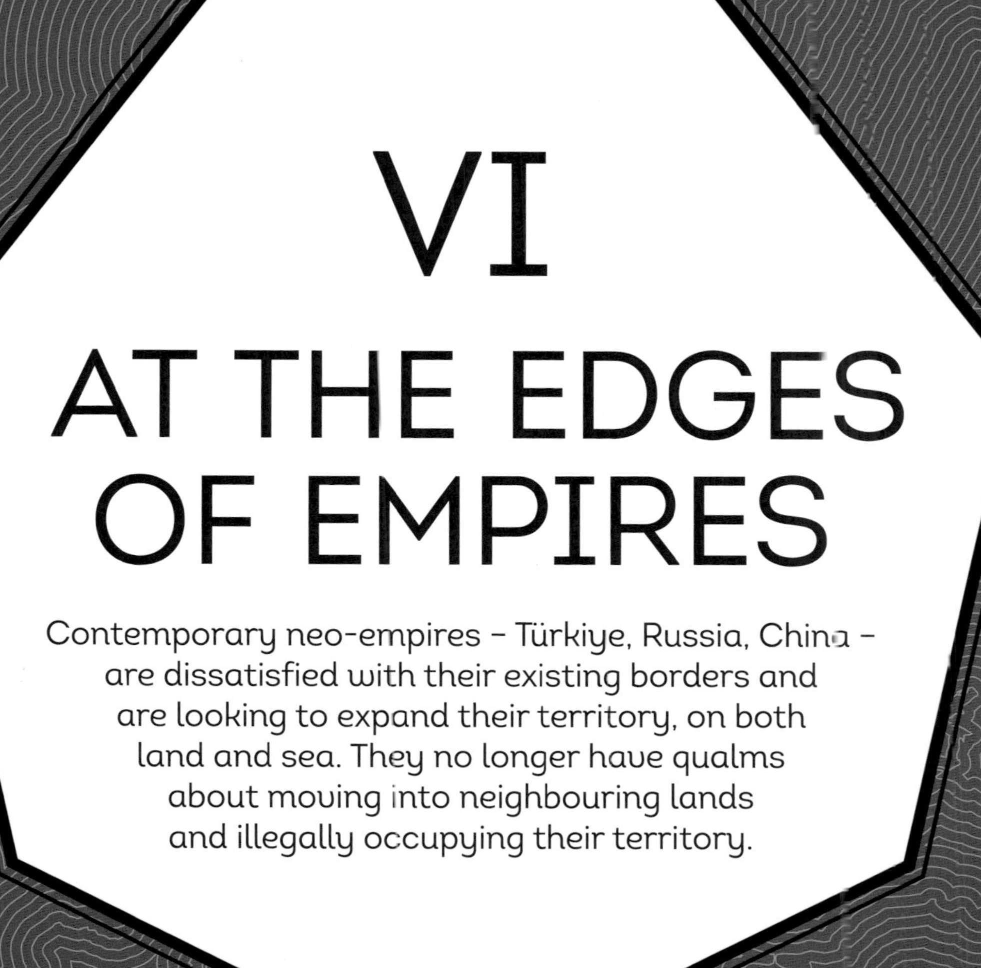

VI

AT THE EDGES OF EMPIRES

Contemporary neo-empires – Türkiye, Russia, China – are dissatisfied with their existing borders and are looking to expand their territory, on both land and sea. They no longer have qualms about moving into neighbouring lands and illegally occupying their territory.

VI · AT THE EDGES OF EMPIRES

Since the early 2010s, President Erdoğan of Türkiye has increasingly spoken of the nation's history as the heart of the former Ottoman Empire and claimed that Türkiye's 'borders of the heart' are much larger than its actual borders. In doing so, he has called into question the national borders drawn up in the first half of the 20th century. Turkish forces are deployed in Syria, Libya, and in Azerbaijan on the border with Armenia, provoking its European neighbours in the Aegean Sea and the Eastern Mediterranean. It also has a military presence in the Red Sea.

Antagonists during the Cold War after the Sino–Soviet split, Russia and China almost went to war during the 1969 crisis, after a serious border incident on Zhenbao Island. Their long shared border has been the subject of a number of disputes. The two countries resolved their border disputes in 2006.

For many years now, the two countries have been increasingly active, in political, economic and military terms, beyond their recognized borders. For them, it is a question of expanding their sphere of influence into their immediate surroundings. This may well have its origins in their deeply rooted identity as empires rather than nation-states.

When it gained independence, Russia was tempted to redefine the country's 'natural' borders by adding Crimea and northern Kazakhstan – but the fate of the former Yugoslavia dissuaded Boris Yeltsin. Since 1991, Moscow has sought to unite most of the former Soviet republics around it. The creation of the Commonwealth of Independent States (CIS) in 1991 was followed in 1992 by interventions in Moldova (Transnistria), Tajikistan and South Ossetia, under the aegis of the CIS. Operations to 're-establish order' on the borders of Russian territory (Chechnya: 1994, 1999) were born of the same logic. In 1999, Russia also briefly intervened on former Yugoslavian territory (Pristina Airport) during the war conducted by NATO.

Since 2000, Vladimir Putin has been seeking to expand his country's territory. The Collective Security Treaty Organization (2002, six countries), formally connected to the CIS, is in fact a military extension of Russia. The Eurasian Economic Community (2000) was replaced by the Eurasian Economic Union (2015, five countries). After an attack on Estonian government servers attributed to Russia (2007), interventions in Georgia (2008) and Ukraine (2013) ushered in an era of more aggressive military action. The Ukrainian crisis broke with the status quo of European borders defined during the Cold War; an event that was all the more serious because it was accompanied by Russia's undemocratic annexation of Crimea. Since the invasion in 2022, Russia's southeast border has been littered with quasi-states embroiled in frozen conflicts: the 'People's Republics' of Luhansk and Donetsk ('New Russia'), Crimea, Transnistria, Abkhazia and South Ossetia, Nagorno-Karabakh. But Ukraine was getting in the way of Putin's plan to unite Russian-speaking people as part of a greater 'Russian world'. After all, Putin has said that the country's borders lie wherever there are Russian people, and Ukraine's growing ties to the West were unacceptable to Moscow. Putin believes he achieved the irreversible by occupying four Ukrainian oblasts in 2022.

Russia has occasionally shown its pragmatic side, for example, when agreeing the delimitation of its borders with Norway (2010) and Estonia (2014). But it is making moves into the Arctic, claiming the maximum extension of its exclusive economic zone there. It is stepping up patrols in the Baltic Sea and the North Sea, as well as in the Sea of Japan, showing no qualms about violating sovereign spaces that belong to neighbouring countries. At the same time, it has refused to negotiate with Japan over the Kuril Islands, which have been partly administered by Russia since 1945 (Japan claims the Habomai Islands and three others). Finally, Russia demonstrated its resurgence in spectacular fashion by intervening, far beyond its borders, in support of the Syrian regime.

Since the 1980s, China has been seeking to stabilize its borders, for both political and economic reasons. It has resolved the majority of its 23 (!) land border disputes with its neighbours, many of which went back a long way. The solution sometimes involved buying the disputed pieces of land, from Kazakhstan (1999), Kyrgyzstan (2004) and Tajikistan (2011). There are, however, still ongoing disputes over territories that China considers its own, to such an extent that it marks them as Chinese on official maps. In India, it has claimed 140,000 square kilometres since 1959 (Arunachal Pradesh, in the east), while India, in turn, demands the restitution of Aksai Chin, in the west. Since the 1962 conflict with India, there has only been a 'Line of Actual Control', which China sometimes tries to encroach on, as it did in the summer of 2020. Beijing also has a dispute with Bhutan over the Doklam area, which was revived in 2019. In South East Asia, the People's Republic has claimed the entirety of the South China Sea since 2009, drawing its maritime border in a shape known in Vietnam as the 'cow's tongue' (or the 'nine-dash line' – in fact eleven dashes since 2014, when China added Taiwan). China is pursuing a policy of land reclamation on reefs to expand its territory. Its disputes with Vietnam and the Philippines are especially heated. When it comes to the strategically important Scarborough Shoal, the Permanent Court of Arbitration ruled in favour of the Philippines in 2016. With Japan, the major dispute is the one over the Diaoyu Islands (as they are known by China) or Senkaku Islands (as they are called in Japan): incursions into Japanese waters have been rising since the late 2000s. Finally, Beijing considers the island of Taiwan to be part of its national territory. It has also unilaterally extended its air control zone.

From an economic perspective, since 2013 China has been seeking to develop 'new silk roads' (the Belt and Road Initiative), transport routes intended to improve strategic supply lines and make it easier to export goods to the West. For some, this neo-imperial approach recalls the growth of Venice as a trading power during the Middle Ages.

In addition, the strategy sometimes called the 'String of Pearls' aims to develop port facilities around the region. It is true that China, which is significantly expanding its high-seas military capabilities, now patrols some distance beyond its territorial waters, as far as the Gulf of Aden, as part of the ongoing fight against piracy.

57 TÜRKIYE AND ITS NEIGHBOURS

Flexing its muscles

Since the early 2010s, President Erdoğan of Türkiye has increasingly spoken of the nation's history as the heart of the former Ottoman Empire and claimed that Türkiye's 'borders of the heart' are much larger than its actual borders. In doing so, he has called into question the national borders drawn up in the first half of the 20th century. Turkish forces are deployed in Syria, Libya, and in Azerbaijan on the border with Armenia, provoking its European neighbours in the Aegean Sea and the Eastern Mediterranean. It also has a military presence in the Red Sea.

TÜRKIYE MAKES MOVES AROUND THE MEDITERRANEAN ...

On land:

- Turkish military presence, or pro-Turkish militia
- Military or diplomatic conflict in which Türkiye is involved
- Kurdish population

At sea:

- Theoretical maritime borders in the Mediterranean based on the line of equidistance, but disputed by the Turkish government
- Current Turkish territorial waters
- Maritime cooperation zone negotiated in 2019 between the Turkish and Libyan governments

... IN A REGION THAT IT CONSIDERS ITS OWN ...

- Extent of the Ottoman Empire at its height (1683–1699)
- Maritime zone claimed by Türkiye

... AT THE RISK OF UPSETTING THE GEOPOLITICAL BALANCE

On poor terms with former Western allies

US or NATO base (Türkiye has been a member since 1952)

European Union (Türkiye applied to join in 1987)

Close to conservative Sunni ruling powers

Islamist government close to the Muslim Brotherhood

Libyan government supported by Türkiye

Seizing opportunities with Russia

- TurkStream, Russian–Turkish gas pipeline
- 2019 purchase of the Russian S-400 missile system, a competitor of the US Patriot system

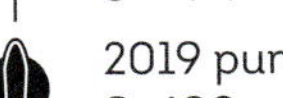
Country where Türkiye has a military presence

As the only NATO member state tha did not impose sanctions on Russia, Türkiye saw its trade with Moscow grow by 42% in the six months following Russia's invasion of Ukrain

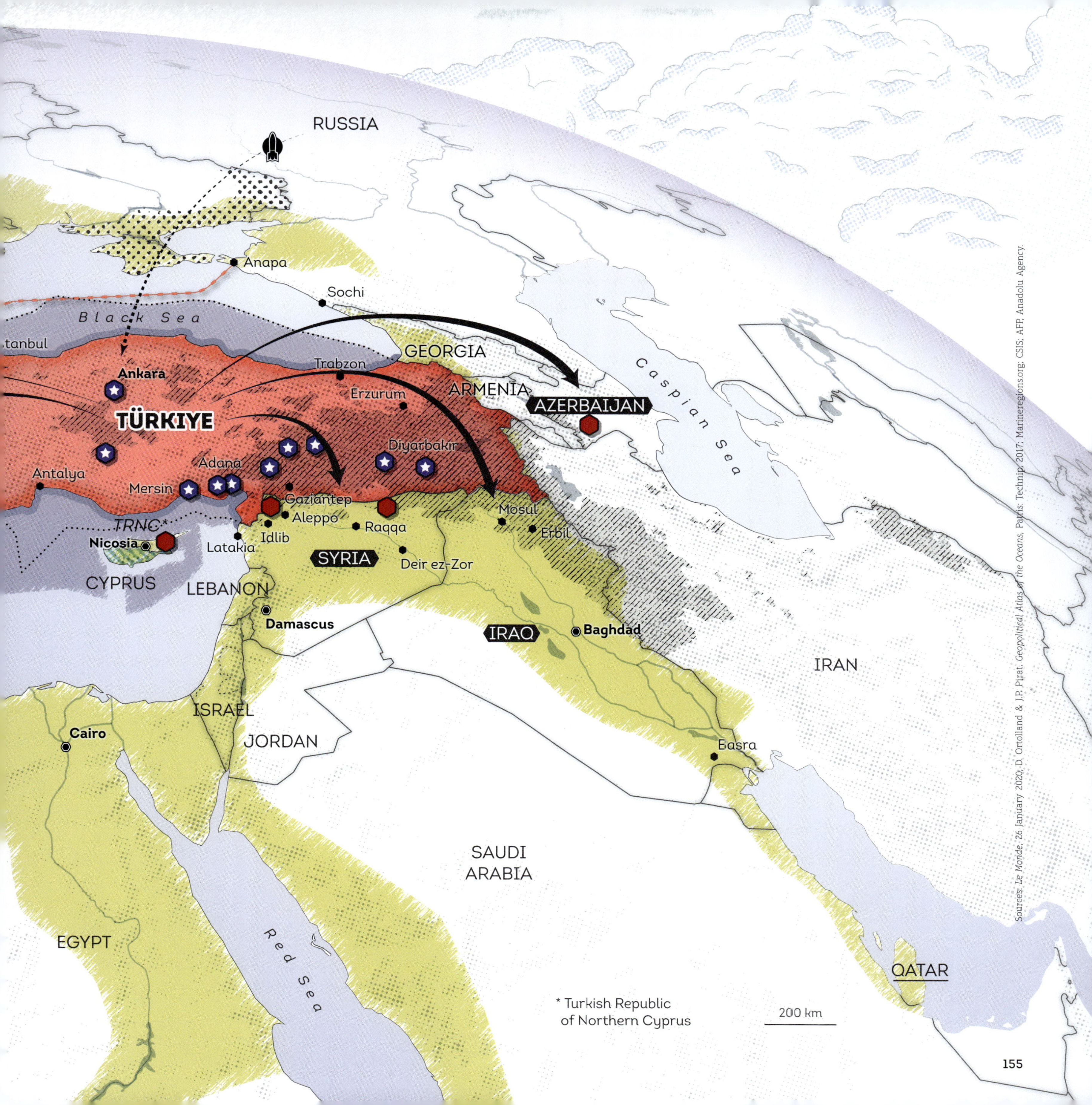

Sources: *Le Monde*, 26 January 2020; D. Ortolland & J.P. Pirat, *Geopolitical Atlas of the Oceans*, Paris: Technip, 2017; Marineregions.org; CSIS; AFP, Anadolu Agency.

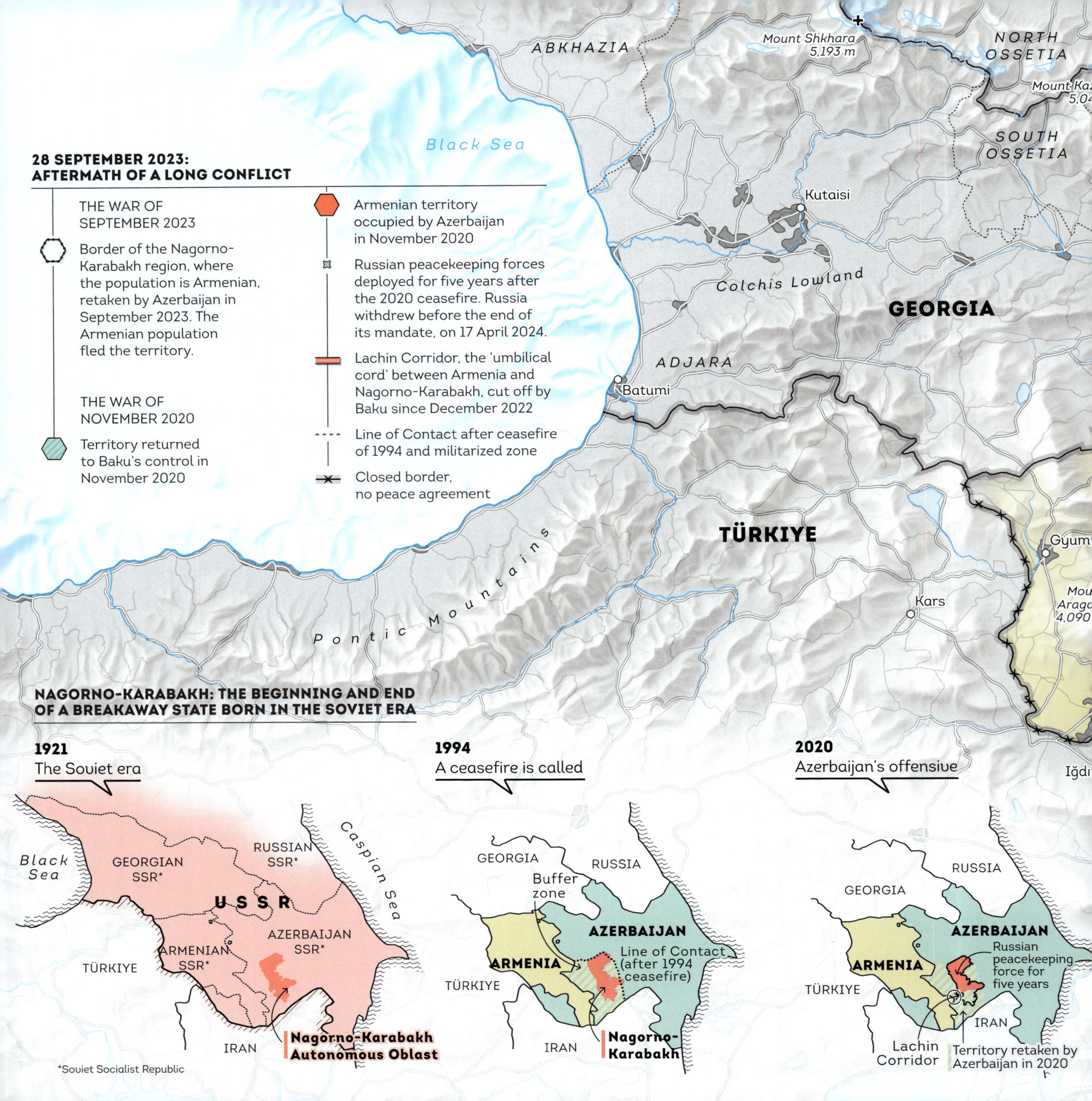

28 SEPTEMBER 2023:
AFTERMATH OF A LONG CONFLICT
THE WAR OF SEPTEMBER 2023
Border of the Nagorno-Karabakh region, where the population is Armenian, retaken by Azerbaijan in September 2023. The Armenian population fled the territory.
THE WAR OF NOVEMBER 2020
Territory returned to Baku's control in November 2020
Armenian territory occupied by Azerbaijan in November 2020
Russian peacekeeping forces deployed for five years after the 2020 ceasefire. Russia withdrew before the end of its mandate, on 17 April 2024.
Lachin Corridor, the 'umbilical cord' between Armenia and Nagorno-Karabakh, cut off by Baku since December 2022
Line of Contact after ceasefire of 1994 and militarized zone
Closed border, no peace agreement
ABKHAZIA
Mount Shkhara
5,193 m
NORTH OSSETIA
SOUTH OSSETIA
Black Sea
Kutaisi
Colchis Lowland
GEORGIA
ADJARA
Batumi
TÜRKIYE
Pontic Mountains
Kars
NAGORNO-KARABAKH: THE BEGINNING AND END OF A BREAKAWAY STATE BORN IN THE SOVIET ERA
1921
The Soviet era
Black Sea
GEORGIAN SSR*
RUSSIAN SSR*
Caspian Sea
USSR
ARMENIAN SSR*
AZERBAIJAN SSR*
TÜRKIYE
IRAN
Nagorno-Karabakh Autonomous Oblast
*Soviet Socialist Republic
1994
A ceasefire is called
GEORGIA
RUSSIA
Buffer zone
AZERBAIJAN
ARMENIA
Line of Contact (after 1994 ceasefire)
TÜRKIYE
IRAN
Nagorno-Karabakh
2020
Azerbaijan's offensive
RUSSIA
GEORGIA
AZERBAIJAN
ARMENIA
Russian peacekeeping force for five years
TÜRKIYE
IRAN
Lachin Corridor
Territory retaken by Azerbaijan in 2020

58 NAGORNO-KARABAKH

Rise and fall of a breakaway state

Sources: Emmanuel Grynszpan and Faustine Vincent, 'Vie et mort de la république du Haut-Karabakh', reprinted in *Le Monde*, 30 September 2023; J. Radvanyi, *Atlas géopolitique du Caucase*, Paris: Autrement, 2009; caucasus.liveuamap.com; War Mapper.

A RUSSIAN EMPIRE

From Novgorod to the Soviet Union

NOVGOROD (9TH–13TH CENTURY)

Territory of the Rus' in the 12th century

In around 860 CE, the Varangian prince Rurik founded the principality of Novgorod, bringing together the Slavic people, with Kyiv as its capital, in 882. In 988, Prince Vladimir converted to Christianity and ruled over the land from the Baltic to the Black Sea. In the 13th century, the principality split apart when faced by the Mongol invasion; Kyiv was destroyed. In 1242, Alexander Nevsky halted Sweden and the Teutonic knights' expansion into Russian territory. He became Grand Prince of Novgorod from 1252 to 1263.

Arctic Ocean
Atlantic Ocean
EUROPE
AFRICA
Tallinn
Karelia
Baltic Plains
Warsaw
Minsk
MOSCOW
Vienna
Kazan
Urals
Kyiv
Chișinău
Astana
Kazakh Steppe
Black Sea
Istanbul
Athens
Mediterranean Sea
Tbilisi
Caucasus
Erevan
Baku
Tashkent
Karakum Desert
Ashgabat
Dushanbe
Jerusalem
Tehran

Sources: Mercator Institute for China Studies; CSIS, Asia Maritime Transparency Initiative.

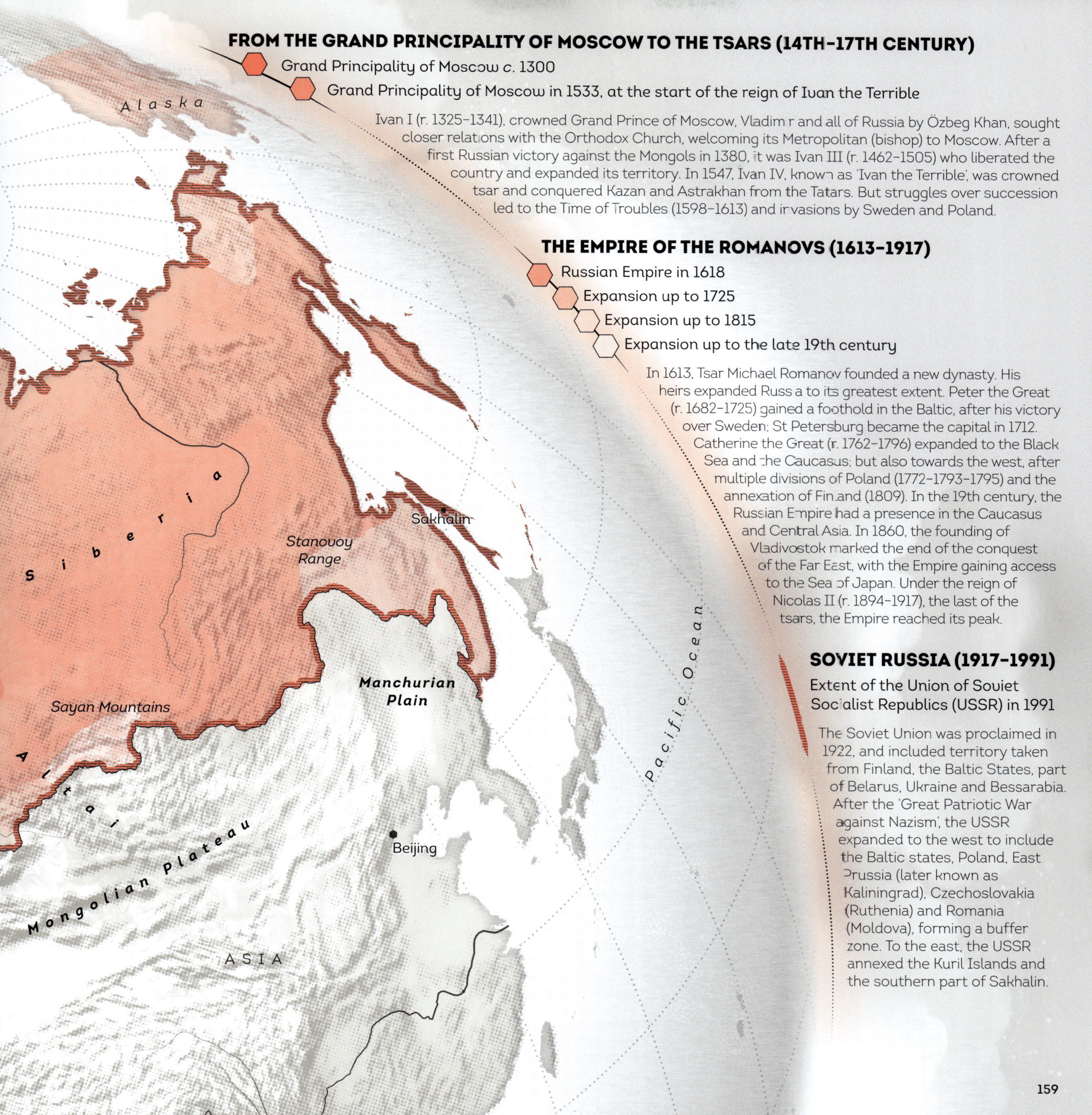

FROM THE GRAND PRINCIPALITY OF MOSCOW TO THE TSARS (14TH–17TH CENTURY)

Grand Principality of Moscow *c.* 1300

Grand Principality of Moscow in 1533, at the start of the reign of Ivan the Terrible

Ivan I (r. 1325–1341), crowned Grand Prince of Moscow, Vladimir and all of Russia by Özbeg Khan, sought closer relations with the Orthodox Church, welcoming its Metropolitan (bishop) to Moscow. After a first Russian victory against the Mongols in 1380, it was Ivan III (r. 1462–1505) who liberated the country and expanded its territory. In 1547, Ivan IV, known as 'Ivan the Terrible', was crowned tsar and conquered Kazan and Astrakhan from the Tatars. But struggles over succession led to the Time of Troubles (1598–1613) and invasions by Sweden and Poland.

THE EMPIRE OF THE ROMANOVS (1613–1917)

Russian Empire in 1618

Expansion up to 1725

Expansion up to 1815

Expansion up to the late 19th century

In 1613, Tsar Michael Romanov founded a new dynasty. His heirs expanded Russia to its greatest extent. Peter the Great (r. 1682–1725) gained a foothold in the Baltic, after his victory over Sweden; St Petersburg became the capital in 1712. Catherine the Great (r. 1762–1796) expanded to the Black Sea and the Caucasus; but also towards the west, after multiple divisions of Poland (1772–1793–1795) and the annexation of Finland (1809). In the 19th century, the Russian Empire had a presence in the Caucasus and Central Asia. In 1860, the founding of Vladivostok marked the end of the conquest of the Far East, with the Empire gaining access to the Sea of Japan. Under the reign of Nicolas II (r. 1894–1917), the last of the tsars, the Empire reached its peak.

SOVIET RUSSIA (1917–1991)

Extent of the Union of Soviet Socialist Republics (USSR) in 1991

The Soviet Union was proclaimed in 1922, and included territory taken from Finland, the Baltic States, part of Belarus, Ukraine and Bessarabia. After the 'Great Patriotic War against Nazism', the USSR expanded to the west to include the Baltic states, Poland, East Prussia (later known as Kaliningrad), Czechoslovakia (Ruthenia) and Romania (Moldova), forming a buffer zone. To the east, the USSR annexed the Kuril Islands and the southern part of Sakhalin.

60 THE EVOLUTION OF UKRAINE

From Kyivan Rus' to 1991

Situated at the edge of an empire, Ukraine grew up alongside its neighbour Russia, with whom it shares large swathes of history. Today, the two nation's conflicting views of the past are used to justify or condemn the Russian invasion of Ukraine.

Borders of the Ukrainian Soviet Socialist Republic in 1922, at the time when the Union of Soviet Socialist Republics (USSR) was first formed.

TERRITORY ACQUIRED...

...in 1939: Polish regions annexed after the German-Soviet Pact

...in 1945: Czechoslovakian region ceded after the Czechoslovak-Soviet Agreement

...in 1948: Romanian regions ceded after the 1947 Paris Treaty

...in 1954: granted by Khrushchev, then First Secretary of the Communist Party of the Soviet Union

TERRITORY CEDED...

...in 1924, to the Russian Soviet Federative Socialist Republic

...in 1940, to the Moldovan Soviet Socialist Republic

Borders of the Ukrainian Soviet Socialist Republic in 1991, when it declared independence

TERRITORY OVER WHICH INDEPENDENT UKRAINE HAS LOST CONTROL...

...after annexation by Russia, following the referendum of March 2014, not recognized by the international community

...after secession of the self-proclaimed People's Republics of Donetsk and Luhansk, annexed by Moscow in February 2022

Sources: *Grand Atlas historique*, Paris: Larousse 2006; *Atlas historique de la Russie*, Paris: Autrement 1997; Marc Ferro (ed.), *L'Etat de toutes les Russies*, Paris: La Découverte, 1993.

61 FALL OF THE IRON CURTAIN

The dismantling of the Soviet Union

Sources: C. Grataloup (ed.), *A History of the World in 500 Maps*, London: Thames & Hudson, 2023; Diploweb.

1922: CREATION

30 December: treaty creating the Union of Soviet Socialist Republics (USSR), which included each republic's 'right freely to secede from the Union' (art. 17)

Borders of the USSR in 1989

Member of the Warsaw Pact (military alliance created in 1955) in 1989

1989: DESATELLIZATION

Satellite state of the USSR that broke away between April and December 1989

2 May: dismantling the Iron Curtain at the border between Austria and Hungary

9 November: East Germany, fall of the Berlin Wall

1990: THE BALTIC STATES DECLARE INDEPENDENCE

LAT. First Soviet republics to declare independence

11 March: Lithuania (ratified in August 1991)

30 March: Estonia and Latvia (ratified in August and September 1991)

1991: COLLAPSE OF THE UNION

11 January: Red Army deployed to the Baltic States

17 March: referendum on the 'preservation of a renewed Union' held by Mikhail Gorbachev, President of the USSR

Republic that boycotted the referendum

83% Proportion of the population in favour of the Union (USSR: 76.4%)

9 April: Georgia declares independance

1 July: dissolution of the **Warsaw Pact**

19–21 August: failed military coup by conservatives opposed to Gorbachev's reforms. Countries start to declare independence:

24 August: **Ukraine**
27 August: **Moldova**
31 August: **Kyrgyzstan**
2 September: **Armenia**
27 October: **Turkmenistan**
25 August: **Belarus**
30 August: **Azerbaijan**
1 September: **Uzbekistan**
9 September: **Tajikistan**

8 December: Minsk Agreement confirming the break-up of the USSR and creating the Commonwealth of Independent States (CIS)

RUS. State that signed the agreement, recognizing the territorial integrity and inviolability of the borders of other signatories

12 December: Russia ratifies the Minsk Agreement

16 December: Kazakhstan declares independence

25 December: Gorbachev resigns

CIS member state

Withdrew later

Associate member state

Pacific Ocean

ASIA

62 RUSSIA'S RESURGENCE

Aiming to increase its sphere of influence

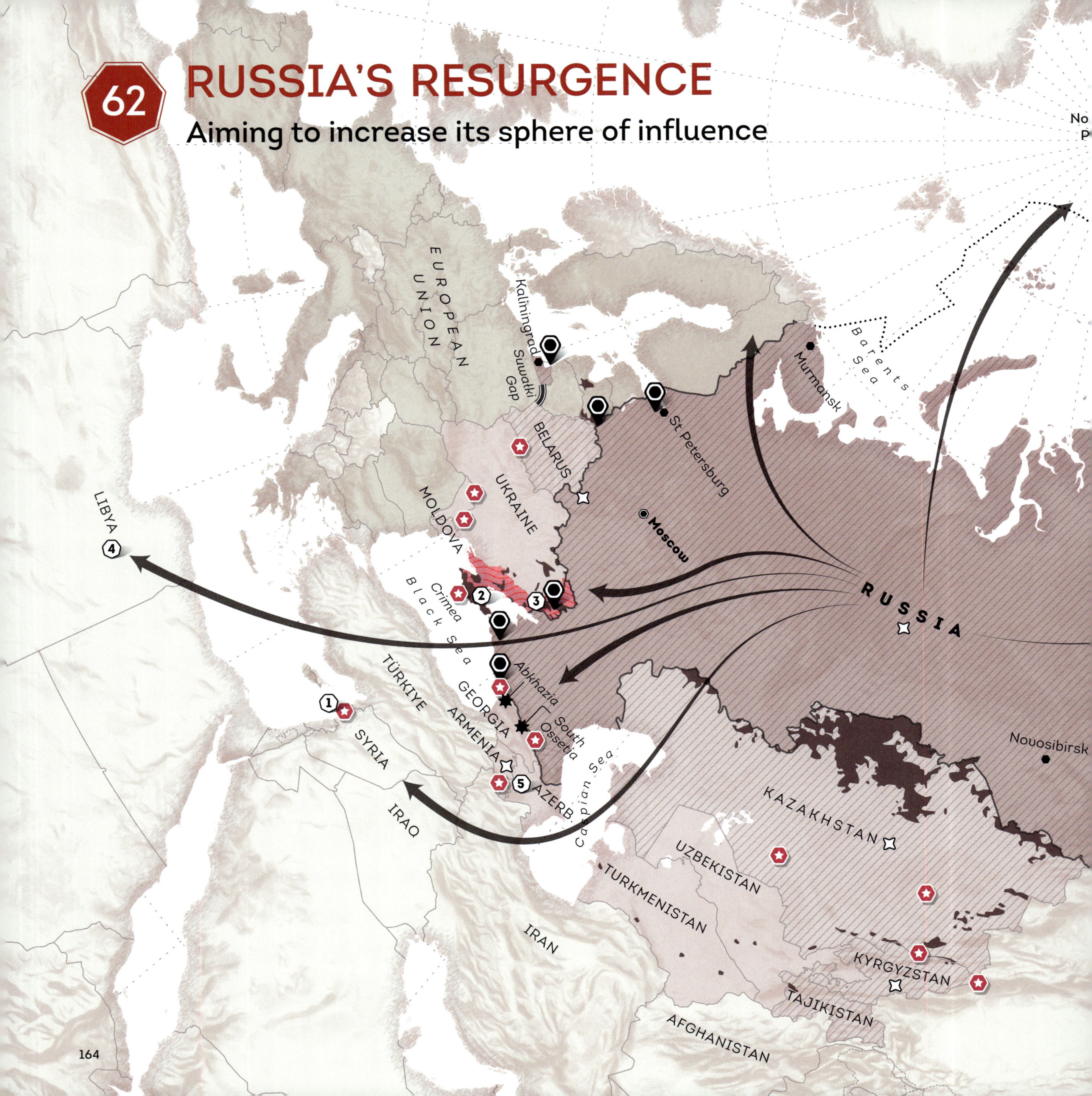

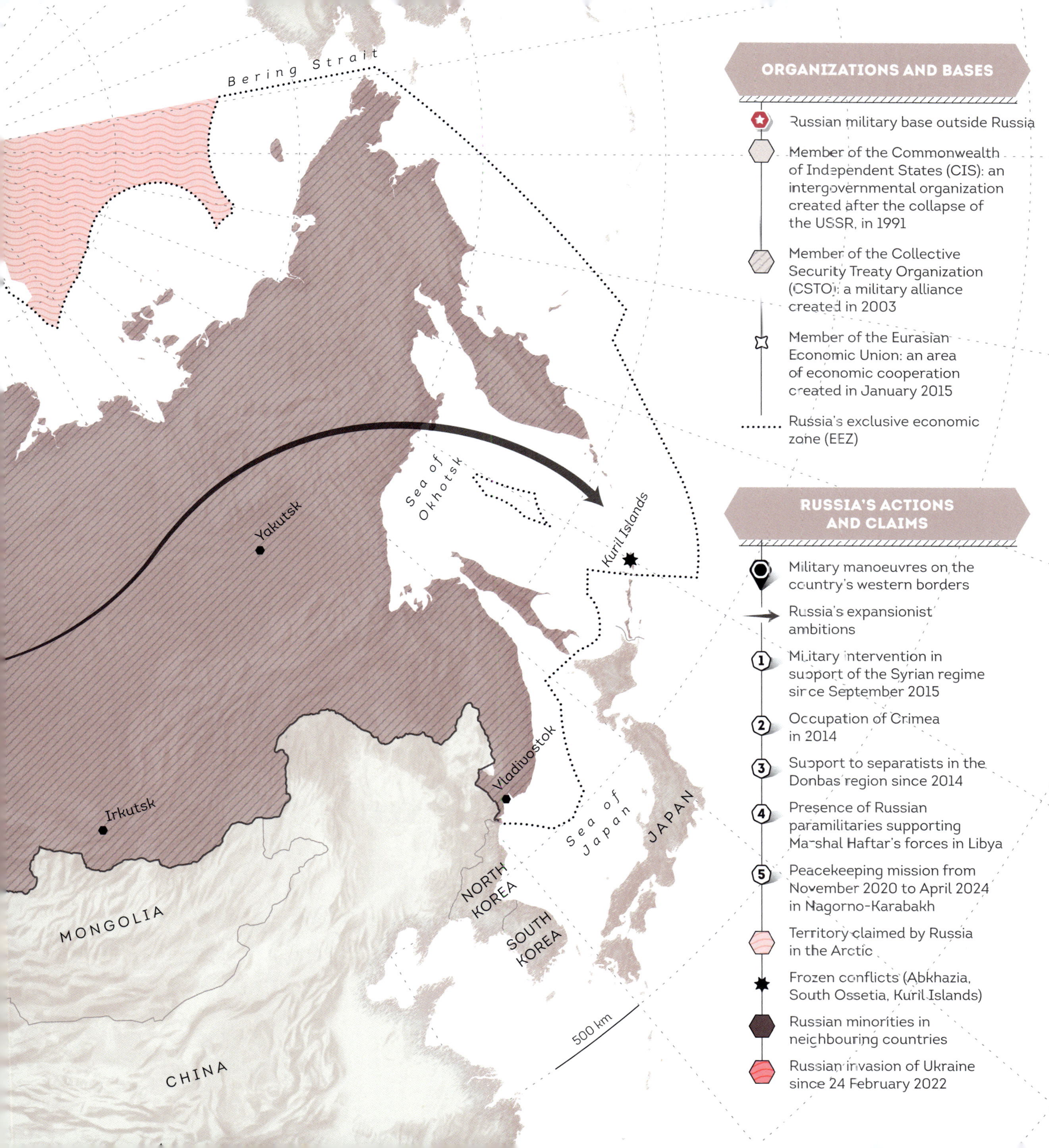

Bering Strait

Sea of Okhotsk

Yakutsk

Kuril Islands

Irkutsk

Vladivostok

Sea of Japan

JAPAN

NORTH KOREA

SOUTH KOREA

MONGOLIA

CHINA

500 km

ORGANIZATIONS AND BASES

- Russian military base outside Russia
- Member of the Commonwealth of Independent States (CIS): an intergovernmental organization created after the collapse of the USSR, in 1991
- Member of the Collective Security Treaty Organization (CSTO): a military alliance created in 2003
- Member of the Eurasian Economic Union: an area of economic cooperation created in January 2015
- Russia's exclusive economic zone (EEZ)

RUSSIA'S ACTIONS AND CLAIMS

- Military manoeuvres on the country's western borders
- Russia's expansionist ambitions
- 1 Military intervention in support of the Syrian regime since September 2015
- 2 Occupation of Crimea in 2014
- 3 Support to separatists in the Donbas region since 2014
- 4 Presence of Russian paramilitaries supporting Marshal Haftar's forces in Libya
- 5 Peacekeeping mission from November 2020 to April 2024 in Nagorno-Karabakh
- Territory claimed by Russia in the Arctic
- Frozen conflicts (Abkhazia, South Ossetia, Kuril Islands)
- Russian minorities in neighbouring countries
- Russian invasion of Ukraine since 24 February 2022

Sources: Pascal Marchand, *Atlas géopolitique de la Russie*, Paris: Autrement, 2015; Jean Sellier, *Atlas des peuples*, Paris: La Découverte, 2014.

63 WAR IN UKRAINE

A battle over borders in Central Europe

RUSSIA, WITH THE SUPPORT OF ITS ALLIES…

- State allied with Russia
- Territory occupied by Russia in 2014, not recognized as such by the international community
- Separatist territory under pro-Russian control
- xxx soldiers — Russian forces positioned near the Ukrainian border before the invasion
- Major Russian base
- Major Ukrainian base

…INVADED UKRAINE ON 24 FEBRUARY 2022

- Frontline
- Encirclement of Ukraine
- Position of Russian forces on 20 February 2022
- Position recorded on 24 February 2022
- Deployment of Iskander ballistic missiles, capable of carrying nuclear warheads
- Naval forces in the Black Sea and the Mediterranean
- First sites targeted (ports, airports, military bases)

BUT THE WAR STALLED IN THE EAST AND ON THE DNIEPER FRONT

- Russian presence in spring 2024
- Furthest point reached by Russian forces since the start of the war

Sources: 'Dans le Sud et l'Est ukrainiens, le conflit s'installe dans la durée,' *Le Monde*, 2 October 2023; Institute for the Study of War; ACLED.

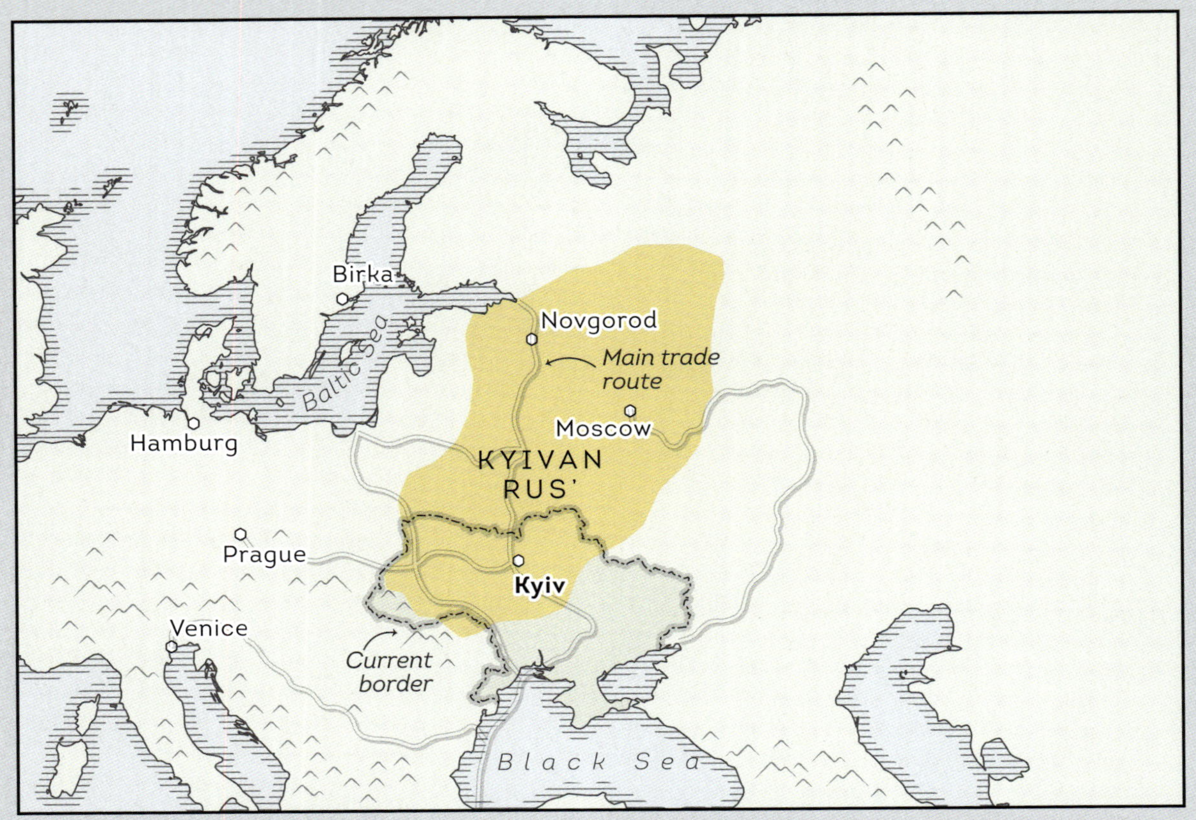

882–1240

KYIVAN RUS', THE FIRST EASTERN SLAVIC CHRISTIAN STATE

In the 8th century, the Scandinavian Vikings (or Varangians) expanded towards the east and opened up a trade route along the river between the Black Sea and the Baltic. Kyiv became an important trading city and a centre of power that brought together the Slavic people under a single ruler. After Grand Prince Vladimir the Great converted to Byzantine Christianity in 988, Kyivan Rus' expanded towards Novgorod and Moscow, then began to decline and broke up after the Mongol invasion.

17th century

UKRAINE DIVIDED BETWEEN RUSSIA AND POLAND

The Cossacks, made up of semi-nomadic people who were joined by peasants fleeing serfdom, spread across the steppe to the borderlands of several kingdoms. The Russian word for borderlands, *ukraina*, became the root of the name 'Ukraine'. In 1654, they became allies of the tsar of Russia. Then in 1667, Russia made an agreement with the Polish-Lithuanian Commonwealth to divide the region between themselves. The territory was split: the land west of the River Dnieper was ruled by Poland, while the land east of the river became Malorossiya, or 'Little Russia'. To the south, the Black Sea coast remained under the control of the Crimean Tatars and the Ottoman Empire.

18th century

RUSSIAN EXPANSION

Catherine the Great expanded the Russian Empire to the steppes and the coast of the Black Sea. Between 1772 and 1795, thanks to the partition of Poland, which had become a Russian protectorate, she reconquered the territory of the former Rus', to the west of the Dnieper, while the Austro-Hungarian Empire acquired the region of Lviv. In the south, she defeated the Ottomans in 1774 and 1783. She therefore controlled Novorossiya ('New Russia'), which stretched from the mouth of the Dnister to the mouth of the Don, taking in the Crimean Peninsula, which provided access to the Black Sea.

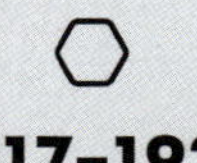

1917–1921

THE REVOLUTIONARY PERIOD

In 1917, the February Revolution broke up the Russian Empire. A Ukrainian parliament, the Rada, was formed in Kyiv and declared independence as the Ukrainian People's Republic. A year later, it signed a separate peace agreement with the central empires, in Brest-Litovsk, alongside the new Russian Soviet Republic. The Ukrainian Soviet Republic, created in Kharkiv in 1918, disputed Kyiv's power and won the civil war that ensued, before joining the USSR in 1922.

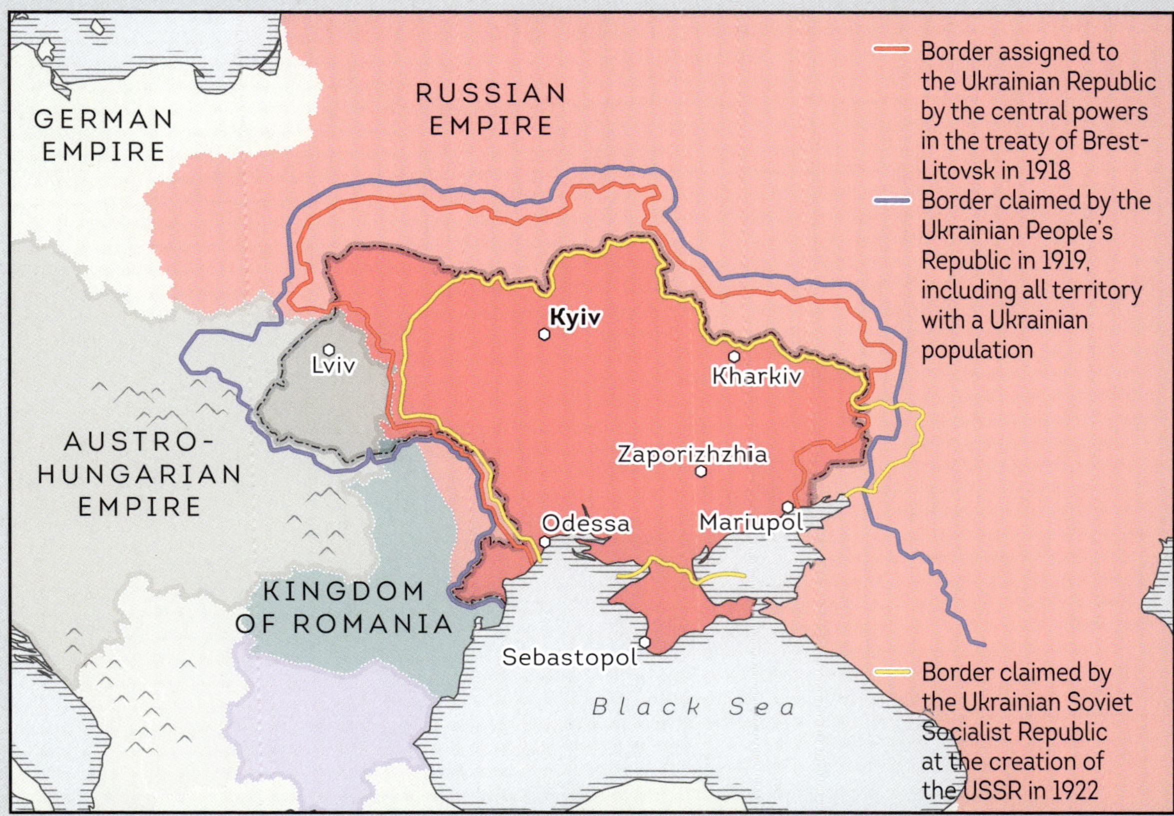

Maps made in collaboration with Thomas Chopard, research fellow at INALCO's Centre de Recherche Europes-Eurasie and deputy director of the Centre d'Études Franco-Russe

64 A NEW IRON CURTAIN?

From Finland to Ukraine

Russia's invasion of Ukraine, along with Belarus's weaponization of migration, have led to a radical shift in attitudes to borders across Europe. Finland and Sweden have joined NATO, making the Baltic Sea part of the West. Norway, Finland, Poland and the Baltic states have reinforced their borders. Stretching for 65 km, the 'Suwałki Gap' (or 'Suwałki Corridor') has become a flashpoint for military tensions. The border between Lithuania and Belarus includes a small panhandle, the Dieveniškės 'appendix', which is sometimes known as 'Stalin's pipe', owing to the legend that Stalin laid his pipe down on the map and cartographers drew the border around it. The Russian exclave of Kaliningrad is served by 'corridor' trains that run through Lithuania. In the south, the land corridor linking Russia, Georgia and Türkiye remains open, as relations between Moscow and Ankara are cordial despite Türkiye's NATO membership.

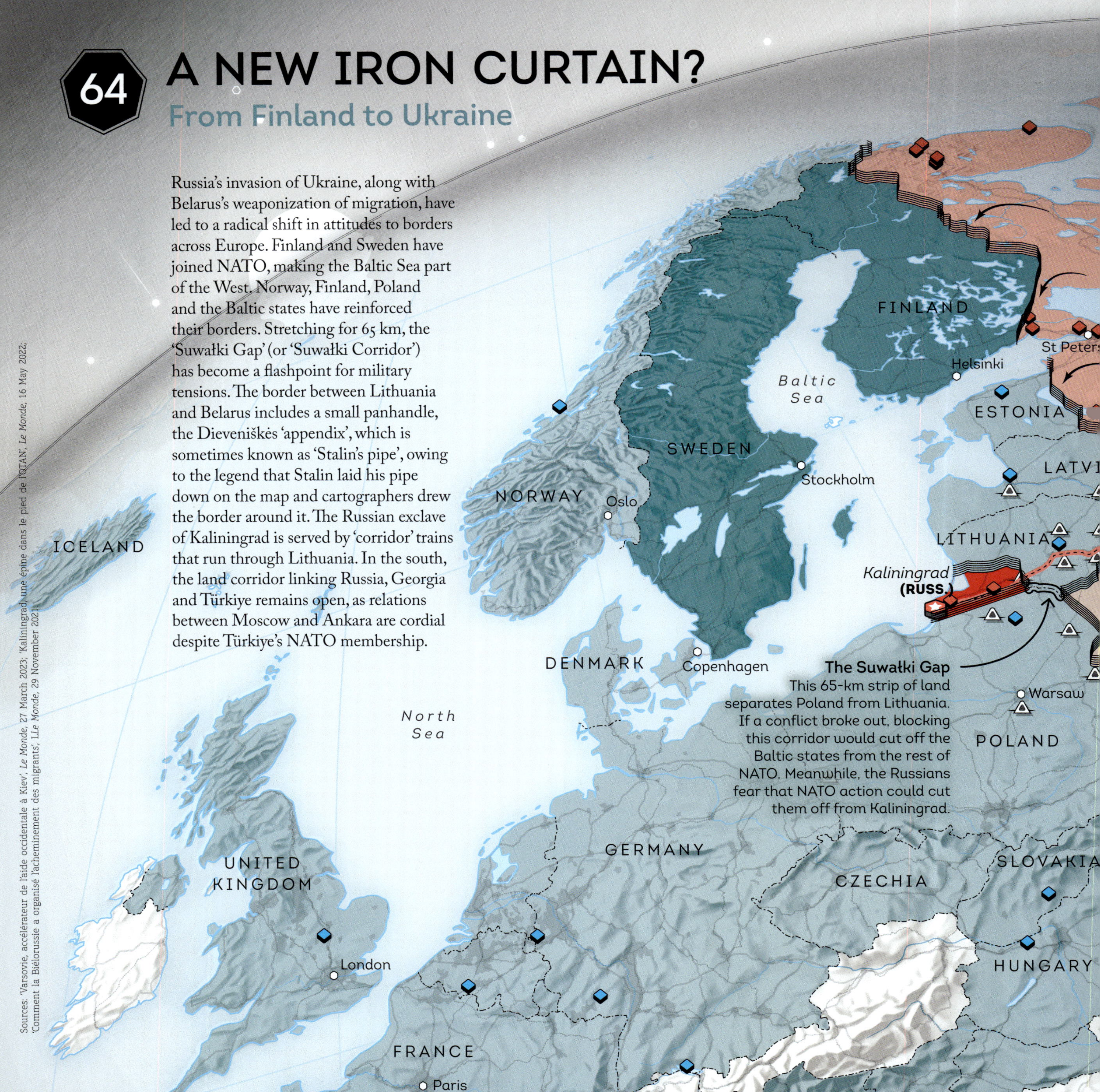

Sources: 'Varsovie, accélérateur de l'aide occidentale à Kiev', *Le Monde*, 27 March 2023; 'Kaliningrad, une épine dans le pied de l'OTAN', *Le Monde*, 16 May 2022; 'Comment la Biélorussie a organisé l'acheminement des migrants', *LLe Monde*, 29 November 2021

Since Russia's invasion of Ukraine...
Russia and its allies
Territory occupied by Russia in 2022
Territory occupied by Russia in 2014
Pro-Russian separatist territory
Russian military bases
Moscow-Kaliningrad railway
Strategically important Russian enclave, nuclear outpost on the Baltic
HQ of the Russian fleet in the Baltic
NATO's eastern flank is building walls and fences
NATO member states
NATO member state that joined after the Russian invasion of Ukraine
Wall or fence, already built or planned
NATO command centres
Remilitarization of the Swedish island of Gotland
Dealing with the influx of migrants being weaponized by Russia and its ally Belarus
Influx of migrants
Major reception centres and detention centres for migrants
A Russian rail line connects Kaliningrad to Moscow via Vilnius. Around a hundred passenger and freight trains travel this route every month, a condition that Russia imposed on Lithuania when it joined the EU in 2004.
MOSCOW
RUSSIA
Minsk
BELARUS
Kyiv
UKRAINE
Area occupied by Russia in 2022
Transnistria
MOLDOVA
Crimea occupied by Russia since 2014
Black Sea
Abkhazia
South Ossetia
GEORGIA
ROMANIA
BULGARIA
TÜRKIYE

65 CHINA AND THE WORLD

Expanding its web of global influence

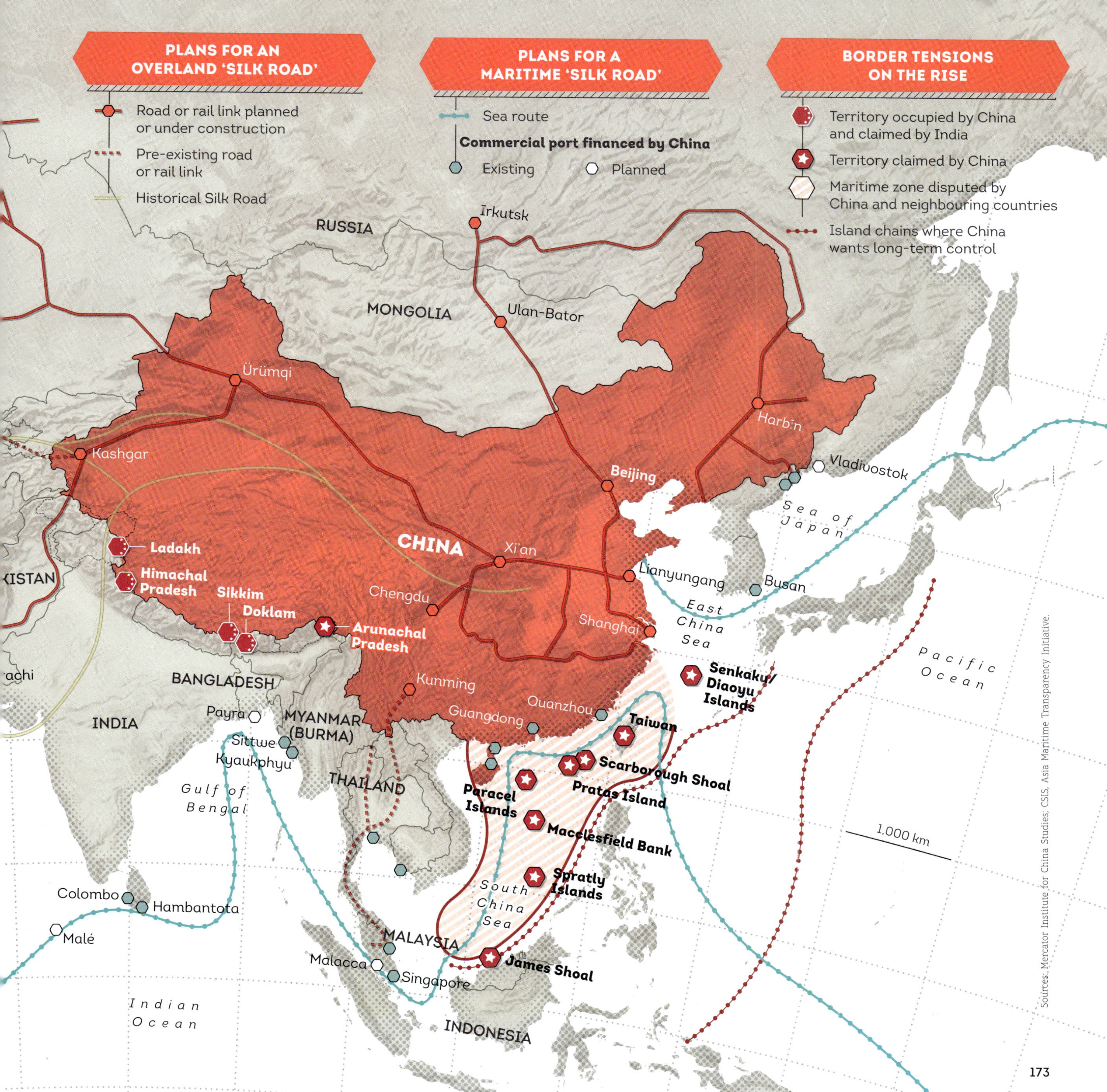
PLANS FOR AN OVERLAND 'SILK ROAD'
Road or rail link planned or under construction
Pre-existing road or rail link
Historical Silk Road
PLANS FOR A MARITIME 'SILK ROAD'
Sea route
Commercial port financed by China
Existing
Planned
BORDER TENSIONS ON THE RISE
Territory occupied by China and claimed by India
Territory claimed by China
Maritime zone disputed by China and neighbouring countries
Island chains where China wants long-term control
RUSSIA
Irkutsk
MONGOLIA
Ulan-Bator
Ürümqi
Kashgar
Harbin
Vladivostok
Beijing
Sea of Japan
CHINA
Xi'an
Lianyungang
Busan
Ladakh
Himachal Pradesh
Sikkim
Doklam
Arunachal Pradesh
Chengdu
Shanghai
East China Sea
Pacific Ocean
Senkaku/ Diaoyu Islands
Kunming
Quanzhou
Guangdong
Taiwan
BANGLADESH
INDIA
Payra
MYANMAR (BURMA)
Sittwe
Kyaukphyu
THAILAND
Scarborough Shoal
Pratas Island
Paracel Islands
Macclesfield Bank
Gulf of Bengal
1,000 km
Spratly Islands
South China Sea
Colombo
Hambantota
Malé
MALAYSIA
Malacca
Singapore
James Shoal
Indian Ocean
INDONESIA
Sources: Mercator Institute for China Studies; CSIS, Asia Maritime Transparency Initiative.

66 TAIWAN

Under pressure from China

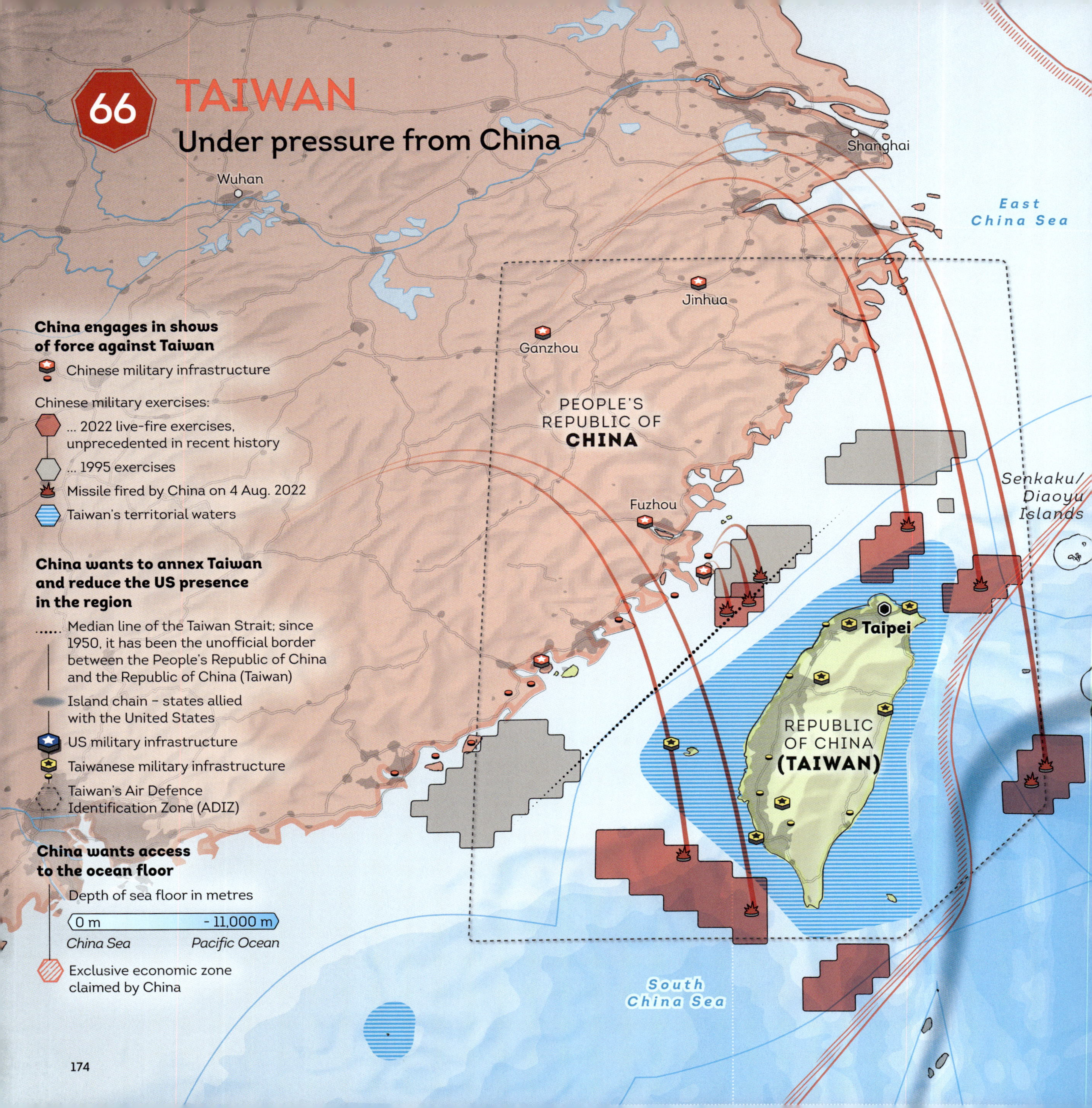

The unique status of Taiwan, or the Republic of China – which has never declared independence and is considered by Beijing to be a province of the People's Republic of China – means there is no official 'international border' between the two countries. In the Taiwan Strait, the median line serves as an unofficial border, but the furthest outlying islands of Taiwan are very close to the Chinese mainland. China regularly tests Taiwan by entering its Air Defence Identification Zone. The number of incursions into the zone is currently rising (many hundreds per month) and they are lasting longer and longer, with planes and ships also crossing the median line, especially during military exercises. China has also started flying drones above Taiwanese airspace. What's more, Beijing has stepped up its military pressure by placing hundreds of ballistic missiles on the coast, within range of the island. Taiwan is also a neighbour of Japan and the Philippines, allies of the United States, but, like Beijing, it claims the Senkaku/Diaoyu islands, as well as a significant part of the South China Sea.

Sources: 'Menaces sur le détroit de Formose', *Le Monde*, 28 November 2022; The Military Balance; ISS; R.O.C. Taiwan Ministry of National Defense.

CONCLUSION

THE FUTURE OF BORDERS

The future of borders is first and foremost the future of nation-states: if, as seems likely, the state remains the most important legal entity in international society, it is not difficult to predict that borders are here to stay, even if they continue to be subject to contradictory pressures – the need to reinforce them on one hand, the desire to cross them on the other.

Since the Limes of ancient times, borders have created a sense of otherness ('them') but also – and for the same reason – identity ('us'). The writer Régis Debray said: 'When we refuse division, is it not a chance to share that we are denying ourselves?' It is a provocative statement, but be that as it may, there is no state, much less a nation-state, that does not have a border. Despite the role played by supranational organizations, transnational businesses and non-state actors, the state remains the main unit in the global system. The construction of post-colonial identities, the weaponization of nationalist sentiment by emerging (or re-emerging) authoritarian powers, and the resurgence of nationalism in Europe and elsewhere all suggest that states, and therefore borders, are here to stay, contrary to what some people hoped or feared in the late 1980s.

What role, precisely, does a border play in the existence of a state? We know the traditional criteria that need to be fulfilled for a state to exist from a legal perspective: a fixed territory, a permanent population, the exercise of sovereignty and international recognition. There is a certain vagueness to these, which explains why we cannot respond with a precise number to the question: 'How many states are there in the world?' There are 193, if we count UN members. But there are more if we add permanent observers – Palestine, the Holy See, non-sovereign entities such as the European Union and the Order of Malta (which has no territory) – and states that are recognized but are not members (Cook Islands, Niue). After that, things become even more complicated.

Palestine is a good example of the complexity of the problem. It is recognized by 143 countries (as of May 2024). But its sovereignty is doubly limited by the quasi-secession of Gaza and the Israeli occupation. Its border is not delimited on the west side (it is simply a ceasefire line and it follows the line – partly within Palestinian territory – of the security barrier). The status of Palestinian territory is a unique case: no devolution of sovereignty from the British colonizers in 1948, Jordan's renunciation of its sovereignty despite the annexation of the West Bank by Amman in 1950; its division into three zones.

Cyprus is a member of the European Union, although the northern part of the island, occupied by Türkiye, has declared itself a 'republic', which is only recognized by Türkiye. Transnistria is an identified territory but it is only recognized as a state by three state entities that are not members of the UN. The Sahrawi Arab Democratic Republic (SADR) has a clearly delimited territory, and is recognized by 45 countries, but the UN classifies Western Sahara as 'non-self-governing' and this situation is unlikely to change in the near future. Taiwan has a well delimited territory but is only recognized as the legitimate government of China by a few countries, and it is not a UN member. The existence

There is no state, much less a nation-state, that does not have a border.

of a territory delimited at least partly by borders is a necessary, but not sufficient, condition for the existence of a state. Neither Somaliland nor New Russia can be considered states. The 'Islamic State', although it had de facto sovereignty over the territories and populations that it administered, had no identified borders, drawing on a long Islamist tradition that called for the revival of the 'Caliphate'. What's more, it was not recognized by any states. As for self-proclaimed micronations, their sovereignty is questionable and they are not recognized by anyone. But future secessions could potentially lead to the establishment of new recognized borders.

The seas are the 'new frontier' and, over the next few years, we will see the delimitation of more maritime territories, applying the UN rulings, but also more disputes in this field, as a number of states seek to argue that they have a physical presence on or around disputed islands. However, there will not be many land borders created out of nothing. As highlighted in the introduction, it is very rare today for such borders to be drawn up. Changing borders by force has become taboo, and the occupation of part of a territory (Golan Heights, Western Sahara, Cyprus, Crimea or Ukraine) is often not recognized on the international stage. Territories seceding or declaring independence often base their borders on existing internal (administrative) borders. The exception may, perhaps, be if the last continent with no borders, Antarctica, were to be one day divided up between neighbouring nations.

It is possible, however, that borders will disappear when territories become unified. As for the strengthening of existing borders, it will continue. There is no reason to think that the trend for building walls and fences will be reversed.

All of which guarantees the future not only of borders, but also of the process that has been christened the 'privatization of borders'. Borders symbolize and give physical form to the state, but their physical management, which is becoming more and more expensive, is often delegated to others. The 'wall industry' is flourishing and is expected to achieve a turnover of 168 billion US dollars in 2025. In airports, tasks previously carried out by the state (air traffic security) have been entrusted to private operators.

Calls for a world beyond borders have been increasing: in the 1970s, with humanitarian movements such as Doctors Without Borders; in the 1980s with greater awareness of environmental dangers (Chernobyl, the ozone layer); in the 1990s with the liberalization of international trade (see Kenichi Ohmae's book *The Borderless World*); in the 2000s with the focus on climate change; in the 2010s with support for migrants (see Europe's No Border network, founded in 1999). All of these problems show a desire for more effective cross-border cooperation. There are growing plans for cross-border sea bridges (Saudi Arabia–Egypt, Bahrain–Qatar, India–Sri Lanka). And that doesn't take into account contemporary threats that leap over or disregard borders, such as long-range missiles, organized crime or cyber-attacks. For the geographer Anne-Laure Amilhat-Szary, 'as these threats do not stop at borders, it is useless to close them'.

As for how to manage borders, this question continues to divide political circles and opinions. Nationalists call for them to be closed, while liberals and alternative-globalists call for them to be opened.

In the future, will we see the elimination of national borders? Or, on the contrary, will they be strengthened? The two trends are not incompatible: freedom of movement supposes that borders are widely accepted and delimited.

There are, in fact, three separate questions: 1) Should borders be open, in the name of tradition, trade or for humanitarian reasons? Or should they be monitored to protect against risks and threats? 2) Should borders be abolished in the name of an ideology, on a regional or global level? Or should they be strengthened in the name of sovereignty? 3) Should we cross borders, or even change them, in the name of national unity or imperialist projects? Or should they be respected and preserved in the name of international stability?

In the long term, we will most likely continue to see clashes between, on one side, the 'Westphalian' world of nation-states and 'sovereignty', and, on the other, the forces of free trade, free movement, pacifism, and messianism (the 'Caliphate'). From an economic perspective, closing borders may be advantageous in the short term (customs duties) but it is disadvantageous in the long term (the positive impact of open borders on growth). Unlike the European

Union, the African Union is now planning to remove its internal borders: there is already a shared passport for members of the Economic Community of West African States (ECOWAS). Some wealthy residents of Silicon Valley are planning to create settlements on the high seas ('sea-steading'), beyond the reach of national governments. But Covid-19 has undoubtedly tipped the balance in favour of the tendency towards strengthening borders.

In Europe, the principle of free movement within the Schengen Area will probably be maintained through strengthening and surveillance of the area's external borders. It has been calculated that ending the Schengen Agreement would lead to a loss of 470 billion euros over ten years for the countries of the EU. Borders in the area could change if states become independent or leave the EU. Likewise, the continuing liberalization of international trade will probably lead to a strengthening of border controls in order to more effectively combat smuggling (people, arms, drugs, plants and animals, works of art), capital flight and tax fraud.

From an economic perspective, closing borders can be advantageous in the short term but it is disadvantageous in the long term.

In the Middle East, it has been fashionable, for some years, to point to the decline of states, the porousness of borders and the power of non-state actors, and therefore to predict that the region will become fragmented. It is true that the situations in Libya and Yemen are ones of de facto partition, a bit like Somalia. But we might equally note the resilience of the external borders drawn up during the long period that stretched from the fall of Ottoman Empire to the end of decolonization, that some of the most porous borders on the planet have recently been strengthened (Saudi Arabia) or are better monitored (Lebanon), and that in many cases (Egypt), the Arab Spring solidified a sense of national identity.

Therefore it would be risky to bet on their disintegration. As Jean-Pierre Filiu said in 2015: 'Over time, these borders have gained legitimacy. They are just as artificial as any others. They delimit the space in which a people's desire for self-determination is exercised. And if they are illegitimate today, it is no longer because of colonization, but because the rights of these peoples have been denied.'

Finally, climate change is already leading to shifts in existing borders and could affect more. The Swiss–Italian border through the Alps has moved by up to 100 metres. The border dispute between Chile and Argentina (Southern Patagonian Ice Field) could be resurrected. Other disputes may disappear: South Talpatti Island emerged from the sea in 1970, and was claimed by both India and Bangladesh, but sank below sea level in 2010. Changes to the courses of rivers could lead to borders being adjusted; one such case occurred between Belgium and the Netherlands in 2015. Rising sea levels will most likely lead to the disappearance of some reefs, which will have consequences for territorial boundaries and disputes between neighbouring nations. In the very long term, entire countries, such as the Maldives, risk the same fate.

67

COVID-19

A virus with no borders

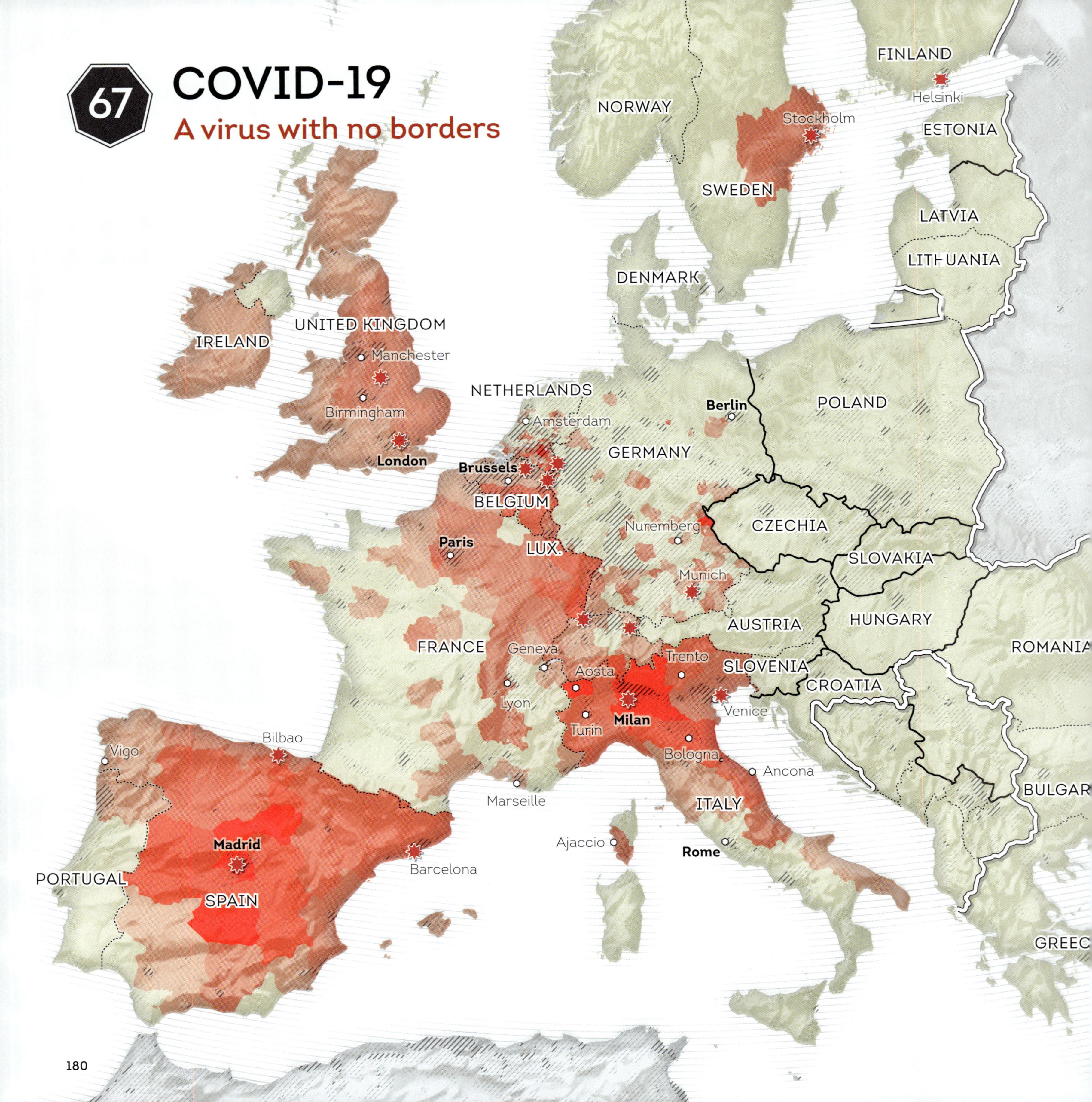

To contain the spread of the Covid-19 virus, which was more concentrated in Europe's major economic centres ...

Death rates linked to Covid-19, per 100,000 people (9 April 2020)

Areas with high concentration of cases

Population density

... European countries imposed travel restrictions

Temporary restrictions on non-essential travel imposed by Brussels on 17 March 2020, applying to all countries in the EU, the four non-EU members of the Schengen Area (Switzerland, Iceland, Liechtenstein, Norway) and the United Kingdom, which did not leave the EU until 31 December 2020

Closed national borders

Increased checks

While Germany managed to slow the spread of the pandemic, the UK and Spain experienced a rising death rate

Number of deaths linked to Covid-19, up to 9 April 2020

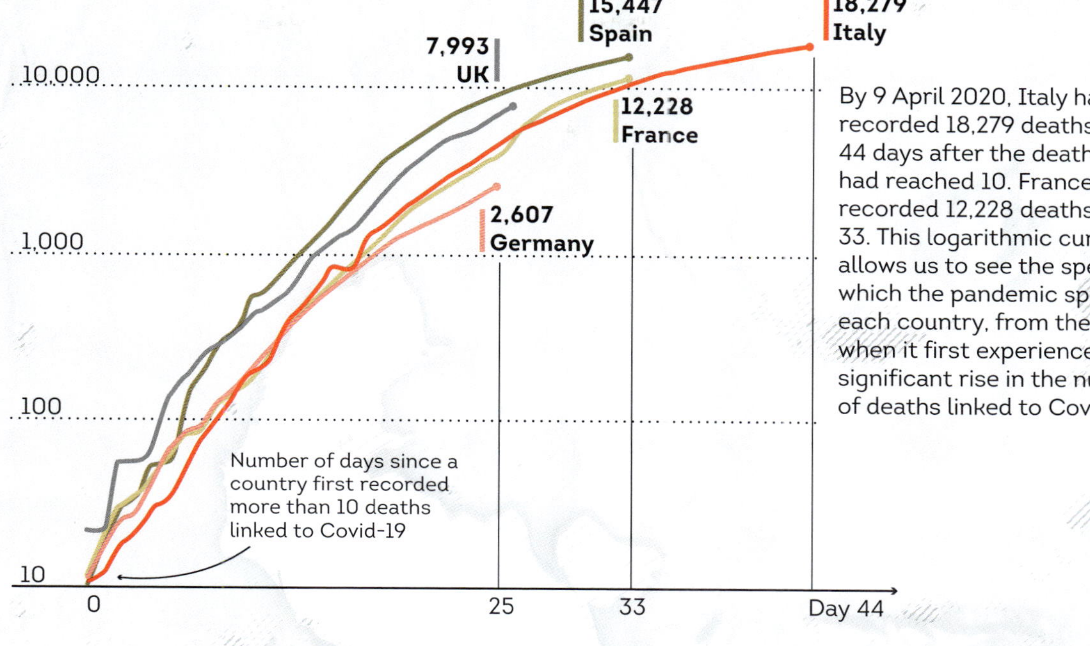

By 9 April 2020, Italy had recorded 18,279 deaths, 44 days after the death toll had reached 10. France had recorded 12,228 deaths by day 33. This logarithmic curve allows us to see the speed at which the pandemic spread in each country, from the day when it first experienced a significant rise in the number of deaths linked to Covid-19.

TESTING RATES

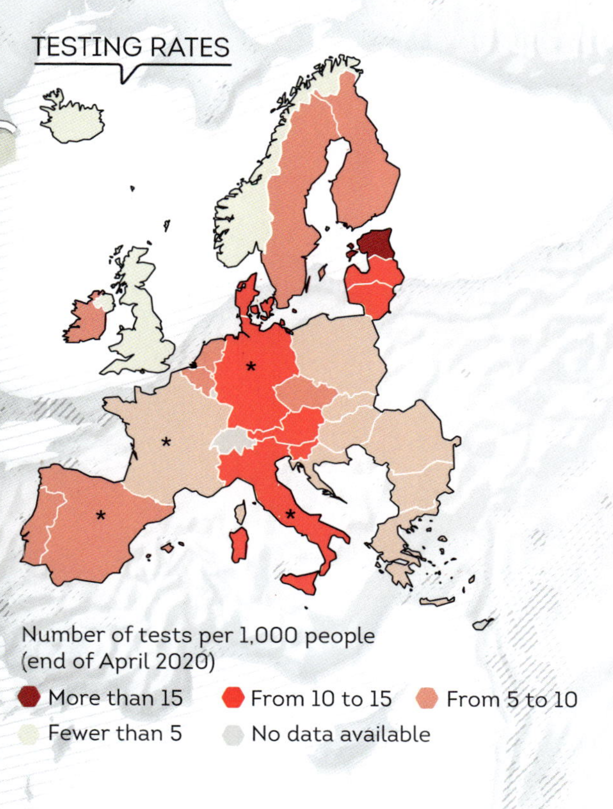

Number of tests per 1,000 people (end of April 2020)

More than 15
From 10 to 15
From 5 to 10
Fewer than 5
No data available

LOCKDOWNS

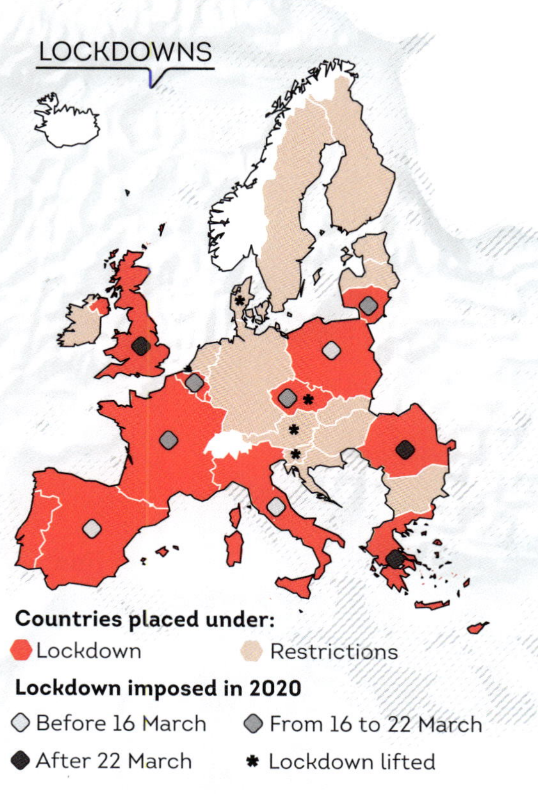

Countries placed under:

Lockdown
Restrictions

Lockdown imposed in 2020

Before 16 March
From 16 to 22 March
After 22 March
Lockdown lifted

PATIENT TRANSFERS

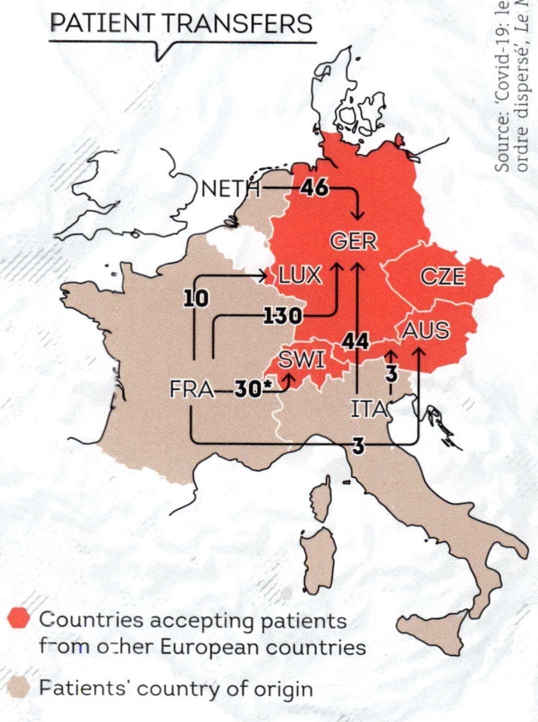

Countries accepting patients from other European countries

Patients' country of origin

Source: 'Covid-19: les Etats européens en ordre dispersé', *Le Monde*, 14 April 2020.

68 A WORLD UNDER LOCKDOWN

Nations in quarantine

Source: 'Sur tous les continents, le confinement à l'origine de déplacements inédits', *Le Monde*, 20 April 2020

4.5 billion people under lockdown ...

Complete lockdown, put in place on this date in 2020

17 March 17 April 25 April 1 May

Partial lockdown, with restrictions limited to specific geographical areas or times of day

Lockdown ending, restrictions lifted

No lockdown imposed

No data

... and millions on the move, right before lockdowns are imposed

- Returning to their country of origin, including repatriations organized by the authorities
- Foreigners leaving (expats, tourists, etc.)
- Expulsion or forced displacement of migrants
- Exodus from the cities (city dwellers moving to the countryside, poorer workers returning to villages, etc.)

Canada
Return of tens of thousands of 'snowbirds', senior citizens spending the winter in Florida

USA
Almost 7,000 detained migrants, from Mexico and Central America, deported within two weeks in early April

Mexico
8,432 Mexicans were able to return between 23 March and 3 April
Many thousands of Guatemalan and Honduran migrants deported

Venezuela
2,135 people allowed to return on 4 and 5 April, and 5,800 more some time afterwards

Argentina
14,950 people repatriated between 18 and 30 March
30,000 Argentinians left the country the day after lockdown was announced

Uruguay
741 people returned from 24 March to 5 April
288 people, mostly Europeans, left the country for Spain

Germany
200,000 people repatriated in three weeks before 6 April

France
From 17 March onwards, 150,000 people repatriated on special flights
Between 1.6 and 1.7 million returned to their home region, and 11% of Parisians left the capital

Spain
Tens of thousands of Madrid residents left the capital on 13 March, the day before lockdown began

Morocco
After flights by Moroccan airlines were suspended on 13 March, more than 28,000 foreigners were repatriated

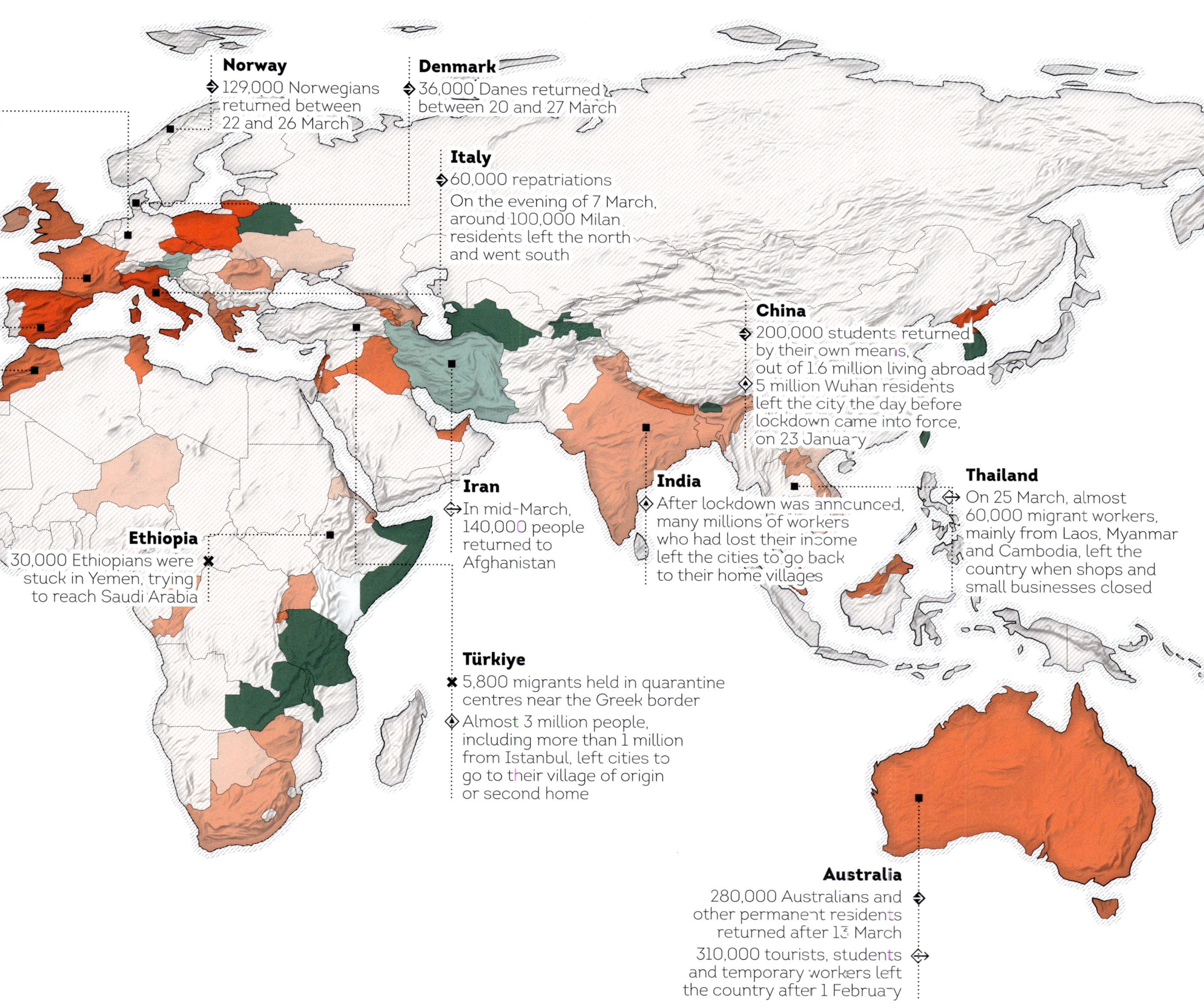
Norway
129,000 Norwegians returned between 22 and 26 March
Denmark
36,000 Danes returned between 20 and 27 March
Italy
60,000 repatriations
On the evening of 7 March, around 100,000 Milan residents left the north and went south
China
200,000 students returned by their own means, out of 1.6 million living abroad
5 million Wuhan residents left the city the day before lockdown came into force, on 23 January
Iran
In mid-March, 140,000 people returned to Afghanistan
India
After lockdown was announced, many millions of workers who had lost their income left the cities to go back to their home villages
Thailand
On 25 March, almost 60,000 migrant workers, mainly from Laos, Myanmar and Cambodia, left the country when shops and small businesses closed
Ethiopia
30,000 Ethiopians were stuck in Yemen, trying to reach Saudi Arabia
Türkiye
5,800 migrants held in quarantine centres near the Greek border
Almost 3 million people, including more than 1 million from Istanbul, left cities to go to their village of origin or second home
Australia
280,000 Australians and other permanent residents returned after 13 March
310,000 tourists, students and temporary workers left the country after 1 February

69 THE FUTURE OF EUROPE?

Dreams of independence

EUROPE IN PIECES: POTENTIAL INDEPENDENT NATIONS

- Member state of the European Union on 1 January 2024
- Border claimed by independence movements
- Regions with the most active independence movements

1: VALLE D'AOSTA
2: SOUTH TYROL
3: VENETO
4: FRIULI
5: ISTRIA

SÁPMI (SÁMI HOMELANDS)
FINLAND
ÅLAND
SWEDEN
ESTONIA
LATVIA
LITHUANIA
North Sea
DENMARK
SCANIA
IRELAND
FRISIA
KASHUBIA
POLAND
NETHERLANDS
FLANDERS
GERMANY
LUSATIA
Atlantic Ocean
WALLONIA
BOHEMIA
MORAVIA
SLOVAKIA
LUX.
BRITTANY
FRANCE
ALSACE
AUSTRIA
HUNGARY
SZÉKELY LAND
2
4
SLO
SAVOY
1
3
CROATIA
ROMANIA
Black Sea
5
BASQUE COUNTRY
OCCITANIA
PADANIA
GALICIA
ANDORRA
MONACO
SAN MARINO
BULGARIA
CORSICA
CATALONIA
ITALY
PORTUGAL
CATALAN COUNTRIES
VATICAN CITY
SPAIN
SARDINIA
GREECE
SICILY
CYPRUS
MALTA
Mediterranean Sea
500 km

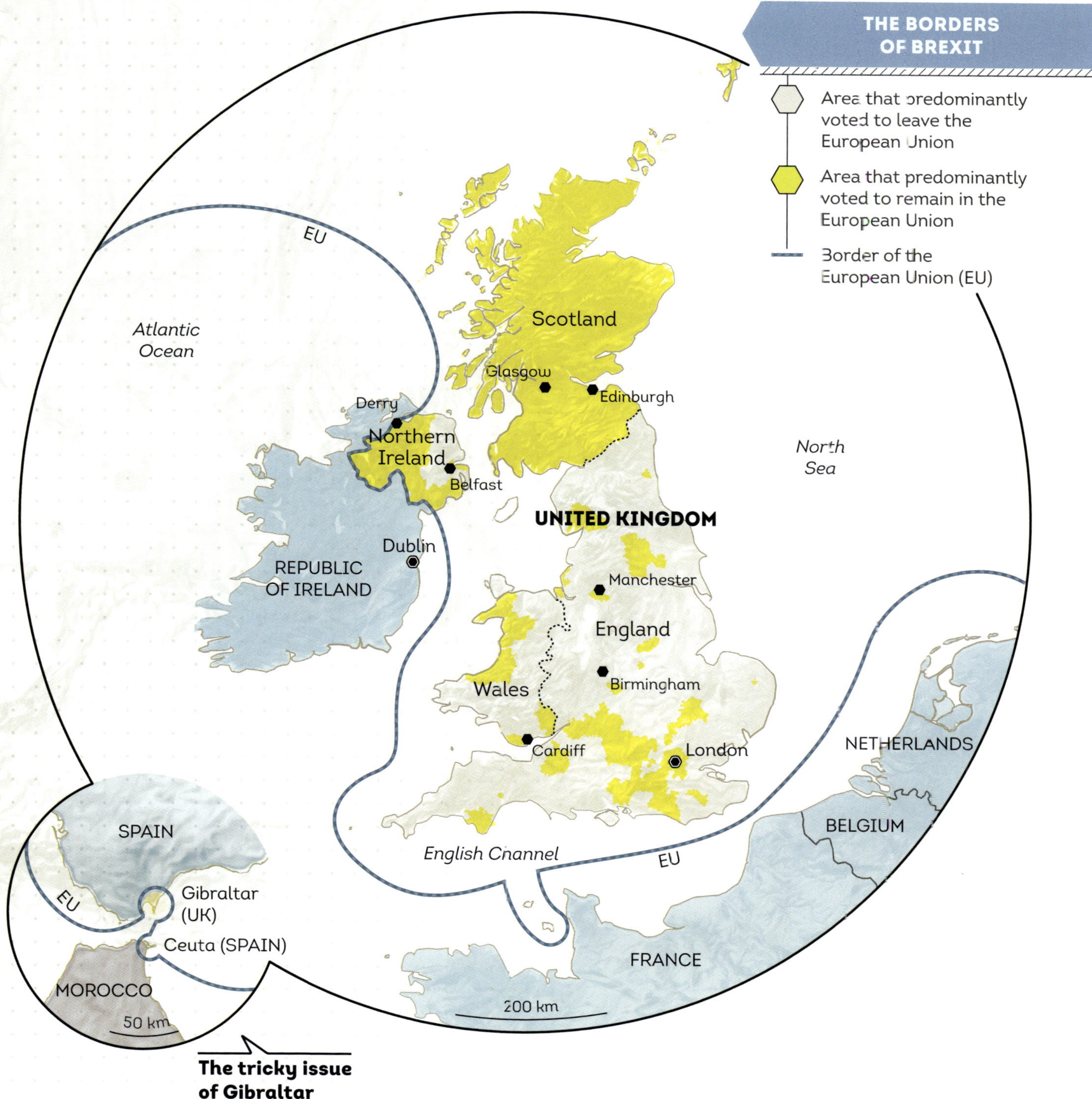

Source: Multinatio.eu

BORDERS IN THE ALPS: THE THREAT OF CLIMATE CHANGE
The Alps is a large region of Europe that overlaps with many borders
Geographic area defined by the Alpine Convention, an international treaty signed by eight countries
Main road routes through the Alps
Border crossing points
National borders in this region are partly determined by natural watersheds
Principal river basins in the Alps
These borders are therefore highly vulnerable to the effects of climate change
Mountain (major peaks and glaciers) where the thawing of the permafrost linked to global warming is making rock walls less stable
Major landslides between 2007 and 2020
BELGIUM
LUXEMBOURG
Cologne
Frankfurt
Stuttgart
Strasbourg
Vosges
Black Forest
Freiburg
Konstanz
Belfort
Basel
Zurich
SWITZERLAND
Besançon
Bern
Jura
Jungfrau 4,158 m
Lausanne
Valais
Ticino
Lake Geneva
Geneva
Matterhorn 4,478 m
Great St Bernard Pass
Simplon
Locarno
Varese
Como
Chamonix
Aosta
Mont Blanc 4,808 m
Savoie
Bourg Saint-Maurice
Chambéry
Milan
Lombardy
Piedmont
Grenoble
Gebroulaz
Fréjus Road Tunnel
Turin
ITALY
Écrins
Briançon
Dauphiné
Col Agnel
Liguria
Genoa
Gap
FRANCE
Rhône Valley
Col de Tende
Gulf of Genoa
Nice
MONACO
Provence
Ligurian Sea
MEDITERRANEAN SEA

70 THE ALPS

Climate change is redrawing borders

In the 1713 Treaty of Utrecht, Italy and Austria adopted the watershed line as the basis for their border. The modern border was defined in 1919 by the Treaty of Saint-Germain-en-Laye and delimited in 1923. In the 1990s, however, it became clear that the Grafferner Glacier was moving, and with it, the watershed line. In 1994, the two countries adopted the principle of 'the moving border'. This took into account shifts that could move the border by many hundreds of metres. This principle was ratified in 2005 for the Italy–Austria border, and used again in 2009 for the border between Italy and Switzerland.

Source: Alpine Convention; World Bank; Alpine Convention for projected changes in annual near surface temperature in Southern Europe 2021–2050; European Environment Agency; Copernixus; ESRI; NOAA; 'Eboulement dans la Maurienne: les axes translapins sous pression', *Le Monde*, 15 September 2023.

GLOSSARY

Terms marked with an asterisk* have their own glossary entry.

Actual Ground Position Line Line of Indian and Pakistani military posts in the Siachen Glacier region. Troops from both countries have been stationed there since 1984.

ADIZ (Air Defence Identification Zone) Region of airspace unilaterally defined by a state, covering the area in which an aircraft must identify itself or be considered an intruder.

Archipelagic state A state made up of a group of islands or archipelago.

Archipelagic waters The waters between and around the islands of an archipelagic state*, which are subject to specific laws under the terms of UNCLOS*.

Artificial border (i) Any border that does not follow natural geographical features. (ii) A border considered to be unjust or illegitimate.

Attila Line Alternative name for the Green Line* in Cyprus, named after Türkiye's 1974 military invasion, Operation Attila.

Balkanization The process of a larger state fragmenting into smaller, often hostile units. The term comes from the fragmentation of the Balkan states in the early 20th century.

Bamboo Curtain The demarcation between Communist and capitalist states in South and East Asia during the Cold War.

Baseline In maritime law, the notional line along the coast from which a state's maritime zones of jurisdiction are measured. It often follows the low tide line.

Berm, The An alternative name for the Moroccan Western Sahara Wall (see map 30).

Blue Line The UN-drawn demarcation line that separates Lebanon from Israel and the Golan Heights.

Borderization The act of installing fences, barriers, surveillance posts, etc. to make the edge of occupied territory resemble a legitimate border.

Buffer zone Neutral area that separates two political or military entities.

Cactus Curtain Nickname for the line dividing the US naval base of Guantánamo Bay from Cuban territory.

Condominium A territory where sovereignty is shared between two or more states.

Contiguous zone Maritime area situated beyond a state's territorial waters, up to a distance of 24 nautical miles, where the coastal state has authority and can enact customs laws.

Continental shelf The part of a continent that lies beneath the ocean's surface. A coastal state has the right to exploit the sea bed and underground resources of the continental shelf, up to a distance of 200 nautical miles.

Delimitation The broad defining of a border between countries.

Demarcation The precise defining of a border between countries.

Desatellization The process of a satellite state* freeing itself from outside control.

Durand Line Demarcation line between Afghanistan and British-ruled India, established by Sir Mortimer Durand in 1893. The border between Afghanistan and Pakistan roughly follows the Durand Line, but Afghanistan disputes its validity.

Dyad A border shared by two neighbouring states.

Enclave A territory belonging to one state but surrounded entirely by another.

Eurosur European border surveillance system.

Exclave A territory belonging to a state but separated from the state's main territory.

Exclusive economic zone (EEZ) an area where a coastal state has exclusive rights to exploit resources, up to a distance of 200 nautical miles from the coast.

Extended continental shelf Maximum area over which the coastal state can have rights over the continental shelf, up to a distance of 350 nautical miles.

Extraterritoriality When part of a state is deemed to exist beyond that state's borders, and is therefore exempt from local laws.

Frontex European border and coastguard agency.

Frozen conflict A situation where armed conflict has stopped but no treaty or agreement has been signed to end the dispute.

Geometric border A border made up of straight lines and arcs, ignoring the physical features of the landscape.

Golden passport Common name for a national scheme that offers citizenship of a particular state in return for investment in that state.

Goldsmid Line Delimitation of the border between Iran and Balochistan, established by General Sir Frederic John Goldsmid in 1872 (modified in 1896 and 1905). Today, it forms part of the border between Iran and Pakistan.

Green Line (1) Ceasefire line between Israeli and Arab forces after the war of 1948–49, drawn by General Moshe Dayan, often wrongly called the '1967 border' (see map 50).

Green Line (2) Ceasefire line between Greek and Turkish forces in Cyprus after the 1975 conflict. Drawn in green pencil on a map of Nicosia by a British officer, it roughly follows the 35th parallel (see map 32) and is also called the Attila Line.

Green Line (3) 1914 delimitation line between the territories of Sindh and Kush (India), now the subject of a border dispute between India and Pakistan.

High seas All maritime areas that are not under the jurisdiction of a specific state; also called 'international waters'.

Ice Curtain The dividing line between the Soviet Union and the USA through the Bering Strait during the Cold War.

Innocent passage A concept defined by UNCLOS*, establishing that ships may pass freely through the territorial waters of another state under a certain set of conditions.

Intangibility of borders Legal principle that consists of respecting existing borders over any unilateral attempt to change them.

Internal waters All of the waters within the baseline* of a coastal state.

International strait Straits separating two or more countries, where ships have the right to free passage.

International waters *see* High seas

Inviolability of borders A key legal principle of a state's international borders – they should not be violated by other states.

Irredentism The concept of a nation reclaiming land that belonged to it at some point in the past.

Iron Curtain The demarcation between Communist and capitalist states in Europe during the Cold War.

Johnson Line Delimitation of the border between Tibet and British-ruled India (1865). This border is now accepted by India but not by China (see map 4).

Kármán line Conventional definition of the edge of space, at an altitude of 100 km above the Earth's surface. Named after the Hungarian physicist Theodore von Kármán.

Limes Latin term referring the outer border of the Roman Empire.

Line of Actual Control (LAC) Ceasefire line between Chinese and Indian forces, after the 1962 Sino–Indian War. Strictly speaking, the LAC only refers to the western portion of the border, in Ladakh. In the east, it merges with the McMahon Line (see map 31).

Line of Contact Ceasefire line established between Armenia and Azerbaijan in the Nagorno-Karabakh conflict (1994–2020).

Line of Control Ceasefire line between Indian and Pakistani forces in Kashmir, 1949 (see map 31).

Macartney–MacDonald Line Delimitation of the border between Tibet and British-ruled India in the west (1893). This border is now accepted by China but not by India (see map 4).

McMahon Line Delimitation of the border between Tibet and British-ruled India in the east, established by Sir Henry McMahon in 1914. This border is not recognized by China, which disputes Indian sovereignty over Arunachal Pradesh. In the east, it merges with the Line of Actual Control (see map 4).

March (also **marches** or **mark**) Archaic term for a borderland or buffer zone*.

Maritime domain The area of sea over which a state has authority. It is divided into areas ranging from closest (territorial waters) to furthest away (exclusive economic zone).

Median line Line of equidistance between two territories.

Micronation A small geographical area that claims to be a sovereign state but lacks any legal recognition.

Military Demarcation Line Frontline between North and South Korea at the time of the 1953 armistice, which now acts as a de facto international border.

Natural border (i) Geographical feature (mountain, river, etc.) used to delimit a border. (ii) Political concept that involves expanding the territory of a state to a border that is considered 'legitimate'.

No-man's-land Unoccupied or neutral ground; often a buffer zone*.

Northern Limit Line Maritime extension (to the west) of the Military Demarcation Line, established by the US, recognized by the UN but not by North Korea (see map 29).

Ohanyan Line Armenian defensive line in Nagorno-Karabakh (1994–2020).

Pemberton Line Line of the India–Myanmar border, established by the British colonizers in 1834.

Plazas de soberanía A set of Spanish overseas territories along the Mediterranean coast of Africa. The Spanish name means 'strongholds of sovereignty'.

Quadripoint A point where four borders meet.

Radcliffe Line Line of the India–Pakistan border (west and east) drawn up by the British authorities in 1947 (see map 4).

Satellite state A country that is officially independent but heavily influenced or controlled by another.

Schengen Area Zone within Europe encompassing 29 countries, with no internal border controls. Named after Schengen, Luxembourg, where the original agreement to establish an area of this kind was signed in 1985.

Securitization An increase in checks and restrictions around a border.

Separation barrier Name given by Israel to the fences and surveillance posts on the western border of the West Bank.

Smart border Border incorporating advanced electronic surveillance and monitoring methods, often in order to speed up transit or improve security.

Sovereignty Defining attribute of a state, referring to its government's ability to act within its territory without external legal or political constraints.

Sykes–Picot Line Line setting out British and French spheres of influence in the Middle East, defined by the accords of the same name (1916), which split the region into five zones. Partially corresponds to the current Syria–Iraq border (see map 3).

Terra nullius Territory that belongs to no state. The Latin term means 'nobody's land'.

Territorial waters Maritime areas over which a state has sovereignty, but where foreign ships have the right to innocent passage*.

Thalweg Line connecting the lowest points of a valley or river bed.

Triple Frontier The tri-border area where Argentina, Brazil and Paraguay meet.

Tripoint A point where three borders meet.

UNCLOS United Nations Convention on the Law of the Sea (1982).

Uti possidetis juris Principle of international law that encourages newly self-governing states to maintain their existing borders, rather than modifying them. The Latin term means 'as [you] possess under the law'.

Watershed line Line identifying the boundary of a drainage basin or water catchment area.

Translated from the French *L'atlas des frontières: murs, migrations, conflits* by Bethany Wright

Cover designed by Steve Russell / aka-designaholic.com

First published in the United Kingdom in 2025 by
Thames & Hudson Ltd, 6–24 Britannia Street, London WC1X 9JD

First published in the United States of America in 2025 by
Thames & Hudson Inc., 500 Fifth Avenue, New York, New York 10110

EU Authorized Representative: Interart S.A.R.L.
19 rue Charles Auray, 93500 Pantin, Paris, France
productsafety@thameshudson.co.uk
www.interart.fr

A CIP catalogue record for this book is available from the British Library

Library of Congress Control Number 2025934129

ISBN 978-0-500-03049-3
01

Printed and bound in Slovenia by DZS-Grafik d.o.o.